Environmental Enrichment of Pigs

Environmental Enrichment of Pigs

Special Issue Editor
Emma Fàbrega i Romans

MDPI • Basel • Beijing • Wuhan • Barcelona • Belgrade

Special Issue Editor
Emma Fàbrega i Romans
IRTA
Spain

Editorial Office
MDPI
St. Alban-Anlage 66
4052 Basel, Switzerland

This is a reprint of articles from the Special Issue published online in the open access journal *Animals* (ISSN 2076-2615) in 2019 (available at: https://www.mdpi.com/journal/animals/special_issues/environmental_enrichment_pigs).

For citation purposes, cite each article independently as indicated on the article page online and as indicated below:

LastName, A.A.; LastName, B.B.; LastName, C.C. Article Title. *Journal Name* **Year**, *Article Number*, Page Range.

ISBN 978-3-03928-078-0 (Pbk)
ISBN 978-3-03928-079-7 (PDF)

© 2020 by the authors. Articles in this book are Open Access and distributed under the Creative Commons Attribution (CC BY) license, which allows users to download, copy and build upon published articles, as long as the author and publisher are properly credited, which ensures maximum dissemination and a wider impact of our publications.
The book as a whole is distributed by MDPI under the terms and conditions of the Creative Commons license CC BY-NC-ND.

Contents

About the Special Issue Editor vii

Preface to "Environmental Enrichment of Pigs" ix

Emma Fàbrega
How Far Are We from Providing Pigs Appropriate Environmental Enrichment?
Reprinted from: *Animals* **2019**, *9*, 721, doi:10.3390/ani9100721 1

Heleen van de Weerd and Sarah Ison
Providing Effective Environmental Enrichment to Pigs: How Far Have We Come?
Reprinted from: *Animals* **2019**, *9*, 254, doi:10.3390/ani9050254 4

Elena Nalon and Nancy De Briyne
Efforts to Ban the Routine Tail Docking of Pigs and to Give Pigs Enrichment Materials via EU Law: Where Do We Stand a Quarter of a Century on?
Reprinted from: *Animals* **2019**, *9*, 132, doi:10.3390/ani9040132 26

Torun Wallgren, Anne Larsen and Stefan Gunnarsson
Tail Posture as an Indicator of Tail Biting in Undocked Finishing Pigs
Reprinted from: *Animals* **2019**, *9*, 18, doi:10.3390/ani9010018 41

Míriam Marcet-Rius, Emma Fàbrega, Alessandro Cozzi, Cécile Bienboire-Frosini, Estelle Descout, Antonio Velarde and Patrick Pageat
Are Tail and Ear Movements Indicators of Emotions in Tail-Docked Pigs in Response to Environmental Enrichment?
Reprinted from: *Animals* **2019**, *9*, 449, doi:10.3390/ani9070449 52

Emma Fàbrega, Míriam Marcet-Rius, Roger Vidal, Damián Escribano, José Joaquín Cerón, Xavier Manteca and Antonio Velarde
The Effects of Environmental Enrichment on the Physiology, Behaviour, Productivity and Meat Quality of Pigs Raised in a Hot Climate
Reprinted from: *Animals* **2019**, *9*, 235, doi:10.3390/ani9050235 64

Lorella Giuliotti, Maria Novella Benvenuti, Alessandro Giannarelli, Chiara Mariti and Angelo Gazzano
Effect of Different Environment Enrichments on Behaviour and Social Interactions in Growing Pigs
Reprinted from: *Animals* **2019**, *9*, 101, doi:10.3390/ani9030101 84

Jean-Michel Beaudoin, Renée Bergeron, Nicolas Devillers and Jean-Paul Laforest
Growing Pigs' Interest in Enrichment Objects with Different Characteristics and Cleanliness
Reprinted from: *Animals* **2019**, *9*, 85, doi:10.3390/ani9030085 93

Jen-Yun Chou, Constance M. V. Drique, Dale A. Sandercock, Rick B. D'Eath and Keelin O'Driscoll
Rearing Undocked Pigs on Fully Slatted Floors Using Multiple Types and Variations of Enrichment
Reprinted from: *Animals* **2019**, *9*, 139, doi:10.3390/ani9040139 106

Nicola Blackie and Megan de Sousa
The Use of Garlic Oil for Olfactory Enrichment Increases the Use of Ropes in Weaned Pigs
Reprinted from: *Animals* **2019**, *9*, 148, doi:10.3390/ani9040148 123

Cyril Roy, Lindsey Lippens, Victoria Kyeiwaa, Yolande M. Seddon, Laurie M. Connor and Jennifer A. Brown
Effects of Enrichment Type, Presentation and Social Status on Enrichment Use and Behaviour of Sows with Electronic Sow Feeding
Reprinted from: *Animals* **2019**, *9*, 369, doi:10.3390/ani9060369 131

Helle Pelant Lahrmann, Julie Fabricius Faustrup, Christian Fink Hansen, Rick B. D'Eath, Jens Peter Nielsen and Björn Forkman
The Effect of Straw, Rope, and Bite-Rite Treatment in Weaner Pens with a Tail Biting Outbreak
Reprinted from: *Animals* **2019**, *9*, 365, doi:10.3390/ani9060365 148

About the Special Issue Editor

Emma Fàbrega i Romans (Ph.D) holds a degree in veterinarian sciences (1995), a master's degree (2000) and Ph.D. (2002) in animal production and a MSc degree in applied animal behaviour and animal welfare (Welfare 1997–1998). Since 2011, she has been a diplomate of the European College of Animal Welfare and Behaviour Medicine. She has worked with the Animal Welfare group since 2003 and has been involved in EU projects (Welfare Quality®, Q-PorkChains, SABRE, Alcasde, IMPRO, ALL-SMART-PIGS, ECO-FCE, FEED-a-GENE and EU PIG). She has coordinated four nationally funded projects on animal welfare: one on alternatives to piglet castration, two on welfare implications of livestock precision farming and welfare implications of livestock precision farming and one on new strategies to improve welfare (free farrowing and environmental enrichment for pigs). Her research topics include assessment of animal welfare on farms, use of precision farming to improve animal welfare, animal welfare in sustainable production systems and alternatives to painful procedures. She is a member of IRTA's Ethical Committee and is the Mediterranean Secretary of the International Society of Applied Ethology (ISAE).

Preface to "Environmental Enrichment of Pigs"

The limitation to fulfil ethological and physiological needs can negatively affect animal welfare and cause the development of abnormal behaviour. Pigs have a strong motivation to perform exploratory and foraging behaviour from a very young age, even if they are provided with enough feed to satisfy their dietary needs. Tail-biting is a redirected behaviour said to be a response to insufficient stimulation and frustration in association with other negative environmental and management factors. Although the exact triggering mechanisms remain unclear, tail-biting has a multifactorial origin, and the scientific evidence has identified a wide range of environmental, dietary and husbandry risk factors. The lack of adequate enrichment material appears to be an initial risk factor that may trigger stress and has been considered a major cause for the most common type of tail biting, the so-called two-stage tail biting, which starts with gentle manipulation of another pig's tail and proceeds to more intensive manipulation with the teeth, causing bleeding and damage. Therefore, tail biting becomes not only an important welfare concern, but it also has serious economic consequences for pig producers, because it lowers weight gains and increases susceptibility to secondary infections, antibiotic use and carcass condemnations. Up to now, tail docking is the most widely preventive measure against tail biting adopted by farmers, although it is considered to cause acute pain, and it does not totally prevent tail biting, because it does not address the underlying causes. The EU legislation of pig welfare does not allow routine tail docking unless other measures such as environmental and management conditions have been tackled. Moreover, EU legislation specifically states that "pigs must have permanent access to a sufficient quantity of material to enable proper investigation and manipulation activities". However, provision of adequate enrichment materials is not easy in modern intensive production systems, because a lot of them use either totally or partially slatted floors. There is a certain mismatch between those materials that are meaningful for pigs to fulfill their behavioural needs (i.e., straw-type materials) and those that do not cause any blockage to the slurry systems (i.e., object-type materials). Therefore, an important challenge for researchers is to deliver data and practical solutions on enrichment materials with attributes that satisfy pigs' needs, that is, being manipulable, investigable, safe, edible and chewable but suitable to be implemented in intensive production systems. The aim of this Special Issue is to collate recent research on "Enrichment Materials for Pigs", tailored to different climatic conditions and pig production systems.

Emma Fàbrega i Romans
Special Issue Editor

Editorial

How Far Are We from Providing Pigs Appropriate Environmental Enrichment?

Emma Fàbrega

Animal Welfare Programme, IRTA, Veïnat de Sies s/n, 17121 Monells, Catalonia, Spain; emma.fabrega@irta.cat

Received: 1 August 2019; Accepted: 17 September 2019; Published: 24 September 2019

Limitations to the fulfilment of ethological and physiological needs can cause countless negative effects on animal welfare and lead to the development of abnormal behaviours. From a very young age, pigs are strongly motivated to perform exploratory and foraging behaviour, even if they are provided with enough feed to satisfy their dietary needs [1]. The widely used term "environmental enrichment" is usually understood as the addition of any element to an environment of captivity. However, from a scientific perspective, the concept of "environmental enrichment" refers to the improvement in the biological functioning of captive animals resulting from modifications to their environment [2]. To provide pigs with adequate enrichment, two main goals should be addressed: (1) the improvement of living conditions to allow the expression of species-specific behaviour, and (2) the development of strategies to manage undesirable behaviours, such as tail biting, and to prevent their escalation.

Tail biting is a redirected behaviour that is said to be a response to insufficient stimulation and frustration in association with other negative environmental and management factors [3]. Although the exact triggering mechanisms remain unclear, tail biting has a multifactorial origin, and scientific evidence has identified a wide range of related environmental, dietary and husbandry risk factors. A lack of adequate enrichment material appears to be the initial risk factor that may trigger stress and has been considered a major cause for the most common type of tail biting, the so-called "two-stage tail biting", which starts with gentle manipulation of another pig's tail and proceeds to more intensive dental manipulation, bleeding and damage. Tail biting is not only an important welfare concern, but also has serious economic consequences for pig producers, because it lowers daily gains and increases susceptibility to secondary infections, antibiotic use and carcass condemnations.

Up to now, tail docking has been the most widely used preventive measure against tail biting adopted by farmers, although it is considered to cause acute pain and it does not totally prevent tail biting, since it does not address the underlying causes. EU legislation on pig welfare does not allow routine tail docking unless other measures such as environmental and management conditions have been tackled. Moreover, EU legislation specifically states that "pigs must have permanent access to a sufficient quantity of material to enable proper investigation and manipulation activities" [4]. However, the provision of adequate enrichment materials is not easy in modern intensive production systems, due to the prevalence of totally or partially slatted floors. There is a certain mismatch between materials that allow pigs to fulfil their behavioural needs (i.e., straw-type materials) and those not causing any blockage to the slurry systems (i.e., object-type materials). Therefore, an important challenge for researchers is to deliver data and practical solutions on enrichment materials that address pigs' needs, that is, manipulable, investigable, safe, edible and chewable, but also suitable for implementation in intensive production systems. Moreover, most of the research on enrichment for pigs has focused on the prevention of tail biting, and therefore, there are gaps of knowledge on how to provide proper enrichment. While studies on the pre-partum period and materials for nest-building purposes have been completed [5], information about sows and piglets during lactation, boars or even gestating sows is scarcer.

Thus, the aim of this Special Issue was to collate recent research on "enrichment materials for pigs", tailored to different climatic conditions and pig production systems.

This Special Issue contains 11 papers related to the enrichment of environments for pigs at different production stages (gestating sows, weaners and fattening pigs). The papers tackle the two previously mentioned main goals of enrichment. Many issues are highlighted in the contributions, and here we only mention a few.

The review paper by Van de Weerd and Ison [6] updates the question of how far we have come in enriching pig environments in the time since the same author (Van de Weerd) published a widely cited review ten years ago [7]. The paper compares the strategies adopted by the three main pig-producing regions (China, US and Western Europe), concluding that, although many improvements have been achieved, we are "still a long way off reaching the ultimate destination of an enriched pig population" (page 16) [6]. A commentary type paper discusses the effectiveness of legislation, especially within the EU context, on enhancing the use of proper enrichment materials to avoid routine tail docking.

In five of the papers, different enrichment materials for fattening pigs are compared, supporting interesting conclusions such as how the interest level of pigs is dependent on the characteristics, presentation, location and maintenance of the objects/materials. More destructible and chewable materials, such as straw in a rack or pieces of wood, were found to be more preferred than less manipulable materials. Tasks such as maintaining the objects' cleanliness or adding new materials were found to be compatible with daily farming routines. One of the studies found a decrease in the interest of pigs over time, and suggested that elements of novelty should be investigated. In another study in which the type of enrichment was varied, tail damage was not reduced, but the decline in pigs' interest towards enrichment was not as pronounced.

Two studies evaluated the effects of enrichment materials at the weaner stage and reported interesting findings, such as the fact that weaner pigs preferred olfactory to non-olfactory enrichment.

One study evaluated three different enrichment materials and presentations for gestating sows and confirmed that following changes in enrichment materials (with a rotation system, for example), sows showed an increased response to enrichment both at the group level and also in a sub-sample of dominant and subordinate focal sows.

One paper compares the effectiveness of relatively small amounts of straw on the floor, a rope or a Bite-Ride during tail-biting outbreaks as intervention measures to reduce the risk of an escalation in tail damage, finding a certain degree of reduction, but also suggesting the need of further research to find more efficient strategies.

Finally, the pig's tail, its posture and/or movement, was used in some of the studies both to predict the likelihood of the occurrence of a tail-biting event (observing an increase in tucked tails), or as an indicator of positive emotional state (i.e., higher tail movement, as indicative of positive emotional state, was found in pigs when interacting with the enriched environment). This final study supports the definition of enrichment material as a way of improving biological functioning and welfare status (i.e., enrichment could enhance positive emotional state) resulting from the modification of a captive environment.

Conflicts of Interest: The authors declare no conflict of interest.

References

1. Stolba, A.; Wood-Gush, D.G. The behaviour of pigs in a semi-natural environment. *Anim. Prod.* **1989**, *48*, 419–425. [CrossRef]
2. Newberry, R.C. Environmental enrichment: Increasing the biological relevance of captive environments. *Appl. Anim. Behav. Sci.* **1995**, *44*, 229–243. [CrossRef]
3. Valros, A.; Munsterhjelm, C.; Hänninen, L.; Kauppinen, T.; Heinonen, M. Managing undocked pigs—On-farm prevention of tail biting and attitudes towards tail biting and docking. *Porc. Health Manag.* **2016**, *2*, 2. [CrossRef] [PubMed]
4. European Union. Council Directive 2008/120/EC of 18 December 2008 laying down minimum standards for the protection of pigs (amending 91/630/EEC). *Off. J. Eur. Union* **2009**, *7*, 5–13.

5. Wischner, D.; Kemper, N.; Krieter, J. Nest-building behaviour in sows and consequences for pig husbandry. *Livest. Sci.* **2009**, *124*, 1–8. [CrossRef]
6. Van de Weerd, H.A.; Ison, S. Providing effective environmental enrichment to pigs: how fare have we come. *Animals* **2019**, *9*, 254. [CrossRef] [PubMed]
7. Van de Weerd, H.A.; Day, J.E.L. A review of environmental enrichment for pigs housed in intensive housing systems. *Appl. Anim. Behav. Sci.* **2009**, *116*, 1–20. [CrossRef]

© 2019 by the author. Licensee MDPI, Basel, Switzerland. This article is an open access article distributed under the terms and conditions of the Creative Commons Attribution (CC BY) license (http://creativecommons.org/licenses/by/4.0/).

Review

Providing Effective Environmental Enrichment to Pigs: How Far Have We Come?

Heleen van de Weerd [1,*] and Sarah Ison [2]

1. Cerebrus Associates Ltd., The White House, 2 Meadrow, Godalming, Surrey GU7 3HN, UK
2. World Animal Protection, 222 Grays Inn Road, 5th Floor, London WC1X 8HB, UK; sarahison@worldanimalprotection.org
* Correspondence: heleen@cerebrus.org

Received: 31 March 2019; Accepted: 15 May 2019; Published: 21 May 2019

Simple Summary: The welfare of farmed pigs can be improved by modifying their environment with bedding, substrates, or objects, so that they can perform more of their pig-specific behaviours. Scientific knowledge on effective enrichment for pigs is not necessarily reaching farms and this paper provides an overview of this issue in the three largest global pork producing regions. In the USA, enrichment has not yet appeared on farms, except when required by higher welfare farm schemes. China hardly has any animal welfare legislation and food safety concerns restrict the use of enrichment on farms. Providing pig enrichment is required by law in EU Member States. In practice, enrichment is not always present, or is unsuitable or inadequate. Other risks to animal welfare include inadequate presentation, location, quantity and size, and maintenance of enrichment. Improvements can be made by applying principles from other fields of behavioural science; welfare knowledge transfer and training to farms; highlighting the economic benefits of effective enrichment; increasing pressure from the financial sector; using novel drivers of change, such as public benchmarking. The poor implementation of scientific knowledge on farms suggests that the industry has not fully embraced the benefits of effective enrichment.

Abstract: Science has defined the characteristics of effective environmental enrichment for pigs. We provide an overview of progress towards the provision of pig enrichment in the three largest global pork producing regions. In the USA, enrichment has not yet featured on the policy agenda, nor appeared on farms, except when required by certain farm assurance schemes. China has very limited legal animal welfare provisions and public awareness of animal welfare is very low. Food safety concerns severely restrict the use of substrates (as enrichment) on farms. Providing enrichment to pigs is a legal requirement in the EU. In practice, enrichment is not present, or simple (point-source) objects are provided which have no enduring value. Other common issues are the provision of non-effective or hazardous objects, inadequate presentation, location, quantity and size or inadequate maintenance of enrichment. Improvements can be made by applying principles from the field of experimental analysis of behaviour to evaluate the effectiveness of enrichment; providing welfare knowledge transfer, including training and advisory services; highlighting the economic benefits of effective enrichment and focusing on return on investment; increasing pressure from the financial sector; using novel drivers of change, such as public business benchmarking. The poor implementation of scientific knowledge on farms suggests that the pig industry has not fully embraced the benefits of effective enrichment and is still a long way off achieving an enriched pig population.

Keywords: environmental enrichment; farming; pigs; sows; welfare; barriers to implementation; USA; China; EU

1. Introduction

The scientific investigation of environmental enrichment originated in the 1960s [1], with a gradual shift in focus from animals in the laboratory, to the zoo, to the farm and to companion animals, as well as a shift in focus on increasing naturalistic behaviour and reducing the incidence of behavioural problems to enhancing welfare.

The term environmental enrichment has been used widely, and the term is often used for anything that gets added to a captive environment. However, from a scientific point of view, it should only be applied to situations where environmental modifications have enhanced the performance of strongly motivated species-specific behaviours or have led to the expression of a more complex behavioural repertoire [2].

Foundational research on environmental enrichment for farmed pigs has focused on the features of housing systems that meet the key behavioural needs of pigs [3]. More recently, applied welfare science has highlighted the main characteristics of effective enrichment. At the same time, awareness of farm animal welfare issues amongst stakeholders such as consumers and food animal producers has increased [4–6].

Furthermore, public pressure for reforms in farming systems also increased because of the risks associated with animal production and international animal trade [7]. Some of these risks have been brought to public attention through high profile animal-health related crises such as BSE (Bovine spongiform encephalopathy), foot-and-mouth disease and African Swine Fever, sometimes coupled to food safety scandals (e.g., clenbuterol fed to pigs in China [8]).

Mellor and Webster [9] describe the process by which societies adapt to increasing knowledge about animal welfare issues as a journey. Countries, regions, businesses and individuals travelling on this journey reach the destination of improved animal welfare at different rates and in different ways. The drivers on this 'journey' will vary in different global regions, but a broad categorisation is possible [10,11]:

- Regulation;
- Concern about animal welfare from consumer/citizens, actioned via non-governmental organisations (NGOs);
- Welfare standards and private animal welfare initiatives (assurance schemes);
- Agri-sector business innovation and business engagement, for example via Corporate Social Responsibility.

Scientific knowledge of animal behaviour and welfare has underpinned progress in each of these categories of driver [12]. While these drivers have been defined for animal welfare in general, they will equally apply to pig welfare.

This paper provides an overview of the progress towards the provision of enrichment that satisfies pig-specific needs ('effective enrichment') in the main global pig-producing regions. This information will illustrate the apparent mismatch between the scientific understanding of what constitutes good and effective enrichment for pigs and the practices on pig farms around the world. Some of the main barriers to providing effective enrichment and possible ways to overcome these will be highlighted and discussed.

2. Science of Pig Enrichment

The main aim of environmental enrichment has been defined as to improve the biological functioning of captive animals [2]. It is therefore important to have detailed knowledge of the particular species concerned when designing and implementing enrichment strategies [13]. Domesticated pigs still express very similar behaviour to their wild ancestors and this has implications for pigs in our farming systems and the behaviours we try to satisfy with the provision of enrichment [14].

The first main goal of enrichment for pigs is to enhance barren living conditions associated with intensive production systems and to provide a suitable outlet for performance of their species-typical behaviour, thus enhancing their welfare. The second goal is to use enrichment as a tool to manage undesirable and damaging behaviours, such as tail biting, to prevent escalation of this behaviour. These two goals are linked as the risk of pig manipulation behaviour is reduced when pigs have a suitable outlet for their species-specific behaviour [15].

In the case of tail biting, enrichment is an important 'emergency' tool, especially when early signs of tail biting are observed (e.g., tail posture [16]). In such situations, additional enrichment (e.g., salt and mineral lick stones, substrates such as Lucerne/alfalfa) can be provided to prevent escalation of the undesirable behaviour, and it is important to apply enrichment that is not used on a daily basis. When novel enrichment is provided, it may reinvigorate the response to and engagement with the enrichment, while animals may have habituated to enrichment that is continually present in the pen [17]. It should be noted that tail biting is a multi-factorial issue [18,19], with many factors contributing to the risks. Enrichment is not the only tool for managing tail biting, and success of the intervention depends on the underlying causes of the specific outbreak.

The prevention of tail biting in growing meat pigs has been the focus of much research on enrichment. These animals are most at risk of this behaviour [1,20]. However, this leaves several knowledge gaps with regards to effective enrichment for other types of pigs. While there is a large body of literature on pre-partum nest building in sows and the use of nesting materials [21], there is much less literature on enrichment for both sows and piglets during the lactation phase. The review by Vanheukelom et al. [22] found beneficial effects on the welfare of both piglets and sows, by providing opportunities to engage in explorative behaviour, nest-building and social interactions and improving maternal responses. These positive effects can also extend into the growing phase [23].

Research on effective enrichment for gestating sows and especially gestating sows in group housing is even scarcer. It is important that these animals are studied separately as their motivation to interact with enrichment is expected to be different from that of growing pigs due to their nutritional status combined with their strong social dominance relationships [24]. Indeed, Stewart et al. [25] found that sham chewing (a stereotypic behaviour that is related to their restricted diets) of pregnant sows in dynamic groups was not reduced by the provision of straw racks, although there was a reduction in pen-directed exploratory behaviour. Aggressive activity may occur at locations of highly valued resources, such as enrichment [26]. Limited access to valued enrichment can result in greater aggression towards subordinate sows and higher stress levels in those animals [27]. Other studies also found increased aggression [25], although, in some studies, aggression remained unchanged after the provision of enrichment [28].

Enrichment for breeding boars has hardly received any attention from the scientific community [1]. These animals are mostly housed individually with limited social contact, and are also (outside Europe), kept in sow stalls with very limited movement and occupation opportunities. The behaviour and housing requirements of mature boars are poorly understood [29] and these animals also require some form of enrichment or a floor covered with a substrate bedding.

2.1. Characteristics of Effective Enrichment

Research has produced a wealth of knowledge on what constitutes effective environmental enrichment [30–32] with associated information on how enrichment can be best applied in practice [18,33]. Functional and effective enrichment needs to meet a series of specific characteristics (Table 1) that will enable pigs to show certain behaviours while interacting with the enrichment. This is linked to the method of provision.

Table 1. The main characteristics of effective pig enrichment (from Van de Weerd [34]).

Main Characteristic	So That Pigs Can ...	Provided in Such a Way That It ...
Investigable	Explore the material with their nose (rooting) and mouth	Remains interesting to a pig (by providing sufficient quantities)
Manipulable	Change the material's location, appearance and structure	Is accessible by suspending it at eye or floor level
Chewable (deformable, destructible)	Manipulate the material by biting and chewing	Is accessible for oral manipulation by all/most pigs in the pen
Edible (with an interesting texture, flavour or smell)	Ingest (eat) the material (that has some nutritional value) (Note: regular feed is not regarded as enrichment)	Is clean, safe and hygienic (minimising the risks of injury or contamination with chemicals or disease-causing agents)

Most types of particulate substrates incorporate the characteristics as listed in Table 1 and are therefore regarded as effective enrichment [35] when provided as bedding (covering the floor area)—and are well maintained—or provided in smaller quantities in racks or dispensers [36].

If bedding cannot be provided (for example due to incompatibility with slatted floor systems or unsuitable geographical climate conditions), producers often resort to providing objects, such as simple chains, plastic pipes or commercially available pig 'toys'. These types of point-source objects (objects limited in size, and restricted to a single location in a pen [1]) may be quick and easy to implement, but their effectiveness is often limited. They may, for example, not prevent tail and ear biting [15,37].

While many studies have evaluated the effect of individual point-source objects, findings are not always consistent or as predicted, due to the range of possible combinations of variables being tested (e.g., method of presentation, location in pen, quantity of material in relation to group size, age and type of pig, etc.) and the confounding of several of these characteristics [17,32].

The main issue with many point-source objects is that the behaviour seen as a result of the interaction is often short-lived and, for pigs, it is mainly intrinsically reinforced (such as exploration). In the absence of a relationship between behaviour and an external consequence such as food, interaction with enrichment is intrinsically motivated [17], so a pig will lose interest following exploration of an object when it has lost its novelty [30,38] and habituation occurs. Conversely, effective enrichment has longer-lasting effects as it provides extrinsic reinforcement. For example, the performance of (intrinsically motivated) exploratory behaviour directed at an enrichment object would be extrinsically reinforced if the behaviour results in finding food items or substrates that can be ingested.

As long as the enrichment continues to provide meaningful (extrinsic) reinforcement, it will retain enrichment value on subsequent exposures [17], for example after an animal has had a period of rest. Rotating enrichment objects, altering the appearance or properties of the items, and increasing the difference between the item and the rest of the animals' environment, as well as providing ingestible rewards on a variable schedule of reinforcement will help to maintain a level of response and slows down the rate of habituation (see [17] for an in depth discussion of intrinsic and extrinsic exploration in the context of enrichment; see also [39]).

Knowledge of the motivation for the performance of a behaviour is very important for predicting the effectiveness of the enrichment. Selecting effective enrichment does not only rely on the characteristics (Table 1) and issues outlined above. There are further aspects to consider for an enrichment programme to be successful. Van de Weerd and Day [1] proposed a framework (building on [2]) that includes health, practical and economic aspects for the evaluation of the success of enrichment. The four criteria are:

1. Enrichment should increase species-specific behaviour;
2. Enrichment should maintain or improve levels of health;
3. Enrichment should improve the economics of the production system;
4. Enrichment should be practical to employ.

Networks led by farmers and the industry can generate practical and effective solutions to animal welfare problems, as they will know the problems to solve and they will know how to design a solution that is meaningful for the specific farming community [40]. However, innovations may be driven by improvements in productivity or profitability, rather than by animal welfare driven and enrichment solutions may not always fulfil all the criteria for effectiveness and can be expensive. Creative producers can make their own enrichment objects by using materials that are easily available on farms (e.g., feed sacks) and by following the principles as outlined in this paper (see also more practical guidance in [34]).

3. Pig Enrichment in Global Practice

Global pork production has increased 4-fold over the last 50 years and is expected to continue growing during the next three decades due to population growth (to 8.6 billion in 2050) and an increased per capita consumption [41].

The top pig-producing regions globally are China (47% of total world production), Western Europe (20%) and the USA (10%) (FAO data from 2016, published in [42]). These three regions represent different stages on the 'animal welfare journey' (see introduction [9]) and, therefore, on their path to enriched pigs. The next section reviews the status of enrichment provision for pigs in these global pork producing regions and describes which of the drivers towards higher welfare play a role locally.

3.1. USA

The USA's 10% share of global pork production [42] accounted for 6% of US agricultural sales (in 2012 (USDA Census Highlights, Hog and Pig farming (accessed on 26 March 2019)). The USA is also the second largest exporter of pork products (FAOSTAT, 2016, Crops and livestock products (accessed on 26 March 2019)). The consumption of pig meat in the USA was decreasing for a short period (2009–2012) but has increased again in recent years and is projected to remain stable with a slight increase in years to come (OECD Meat Consumption data (accessed on 26 March 2019)). These projections may change depending on the development and adoption of alternative proteins (plant-based protein, edible insects, clean or cultured meat) in US diets [41,43].

There is no federal legislation to protect the welfare of farm animals during rearing [44]. Regulations under the Animal Health Protection Act give the Department of Agriculture general authority to broadly protect farm animal health, but not specifically welfare. There is, however, some federal legislation to protect farm animals during transport and slaughter (although poultry are excluded). This includes the 28-h rule for transport to provide rest, feed and water every 28 h, but there is no mechanism for monitoring, meaning enforcement is problematic. The Humane Slaughter Act 1958 dictates that farm animals are handled and slaughtered in a humane way (also excludes poultry) and the Federal Meat Inspection Act provides inspection of handling and slaughter.

At the state level, only 16 of 50 states include livestock in anti-cruelty legislation [45,46]. Thirty states have additional humane slaughter legislation [44]. Considering all animal protection legislation, the top three states are California, Oregon and Massachusetts, with Mississippi, North and South Dakota at the bottom [47]. Through ballot initiatives, 12 states have existing or planned legal limitations on the use of close confinement systems (sow gestation crates, veal crates, and/or battery cages for laying hens), including 10 states limiting the use of gestation crates [45,46]. Additionally, two states (California and Massachusetts) limit the sale of products from close confinement systems. California started this trend in 2010 (Proposition 2, prohibiting the sale of caged eggs), recently strengthened with Proposition 12 (limiting confinement for veal calves, pigs and egg laying poultry).

Pigs are mainly raised on intensive farms. The majority of these are large scale, with two-thirds of pigs housed in farms of over 5000 and 90% on farms with over 2000 pigs (Census of Agriculture United States, 2012 (accessed on 26 March 2019)). Intensive systems typically include total confinement (enclosed, mechanically ventilated buildings) with fully slatted floors and no substrate/bedding or enrichment.

Pig enrichment has not yet featured on the political agenda, nor on the agenda of civil societies. The use of enrichment on farms is sporadic and mainly appears on 'niche farms' and/or those operating under certain third-party audited farm assurance schemes (Animal Welfare Approved; Global Animal Partnership 5-step program (Step 2 and above); American Humane Certified; Certified Humane (all accessed on 26 March 2019)).

'Niche' production for alternative marketing avenues is gaining interest from the US consumer and producer alike [48,49]. These production systems are typically characterised by outdoor and/or bedded systems, heritage breeds, no sub-therapeutic antibiotics or growth promotors, vegetarian feed and a small- to mid-size family farm production setting. They provide a higher standard of welfare for the pigs, a quality product for the ethical consumer with a market premium (often covered by third-party certification schemes) and personal satisfaction for the farmer [48,49].

As part of the 'We Care' initiative, the National Pork Board's voluntary industry-led scheme, Pork Quality Assurance (PQA) Plus (Pork Quality Assurance (PQA) Plus (accessed on 26 March 2019)), aims to promote food safety, public health, a safe working environment and animal well-being. The scheme has two components, individual certification following an education program and site status achieved after farm site assessment by trained assessors. However, the well-being sections of the individual and site assessments exclude the provision of enrichment [50,51].

Interest in the provision of environmental enrichment for pigs in conventional confinement systems currently focuses on its utility to improve production performance. This includes minimising unwanted inter-pig aggression, especially at the point of regrouping, damaging behaviour such as tail, ear and flank biting and stereotypic behaviour [26,52,53]. This interest is in its infancy, meaning peer-reviewed research papers are scarce. However, the National Pork Board's 2019 call for research proposals includes "Review the benefits of enrichment and their impact on preventing aggressive and damaging behaviors in various housing environments and on overall production." In addition, the USDA has environmental enrichment for gestating sows, farrowing sows and piglets and weaner, grower and finisher pigs built into the current work plan [Jeremy Marchant-Forde, Personal communication, 2019]. Iowa State University has a project underway to create and test (in a research and commercial setting) a novel pig dietary enrichment to improve survivability during the weaning transition [Anna Johnson, Personal communication, 2019].

Non-governmental organisations (NGOs) actively campaign on behalf of their supporters to improve animal welfare. These organisations can influence public opinion via activities such as public awareness-raising campaigns and targeted direct action, but also dialogue with industries and regulatory bodies, all with varying levels of impact [54,55]. Combining legislative work with undercover investigations, litigation and corporate engagement, NGOs have played a role in transitioning the pork industry away from gestation crates over the last decade [55]. A significant aspect to this success in the US has been through corporate engagement, raising the gestation crate issue with major pork buyers, including restaurants, grocery stores, fast-food chains, and others in the retail sector [55]. This has now extended to other pig welfare issues including the provision of enrichment (see for an example [56], although reporting shows that only small numbers of pigs receive enrichment and that the provided objects—balls, chains and toys—do not meet all the characteristics of effective enrichment as per Section 2.1). The future outlook is focused on providing pigs with effective enrichment and to better understand the challenges producers face in an uncertain market in order to invest in improved housing conditions for pigs.

A lack of federal legislation protecting farm animal welfare during rearing, as well as few and inconsistent provisions of state legislation, are barriers to improving pig welfare [44]. In addition, concerns for biosecurity and manure management in relation to the use of environmental enrichment [26], as well as a lack of recognition of the need for enrichment by the industry, is likely to further hamper progress. Due to the nature of conventional systems in the US, it is unlikely that enrichment will fulfil the characteristics of effective enrichment without significant infrastructure changes. Consumer-driven improvement in food company animal welfare policy, implementation and monitoring may be the most promising route to achieving effective enrichment in the USA. However, information on public attitudes towards farm animal welfare and perceptions of current and proposed future animal production systems is needed to ensure future systems reflect the public's ethical values about the treatment of animals [57].

3.2. China

China accounts for almost half of the world's pork production and pork was, and still is, the predominant staple meat for Chinese consumers, with around 700 million pigs slaughtered annually—one for every two Chinese people [58]. In addition to being almost the biggest pig producer, China is also the biggest importer of pig meat and shows the biggest increases of imports in the last 10 years [42].

Rural families traditionally raised pigs as part of small-scale farming operations, and these are now increasingly being replaced by larger-scale farms that use grain-based feeds [59]. As a consequence of active government policy (since 2007), the average pork enterprise size has increased towards industrialized production, with large farms taking advantage of scale economies in accessing better genetics, addressing quality issues and mechanizing to replace labour.

There are societal and cultural awareness barriers to improving animal welfare, and a large proportion of Chinese consumers view animal welfare (e.g., improve the rearing conditions) as part of the food safety of animal-derived products [60]. However, there are signs that society is increasingly aware of the concept of animal welfare and of the need to establish laws to improve animal welfare [ibid.] and, more specifically, of the need to improve welfare during transport and slaughter [61]. Animal welfare awareness is higher and more developed in globally-oriented cities such as Beijing and Shanghai with the highest income, education, and greatest exposure to modern lifestyles and information [62].

With regards to animal welfare legislation, the Chinese government has enacted some laws regarding laboratory animals, zoo management and practising veterinarians and these provide some level of general protection to animals. In addition, there is some legislation relevant to farm animals, specifically regarding transport and slaughter. However this is mainly based on food safety concerns that touch on some animal protection [63].

The Chinese Veterinary Medical Association (Chinese Veterinary Medical Association (CVMA, accessed on 4 March 2019), supervised by the ministry of Agriculture, has a branch that deals with animal welfare (Animal Health and Welfare Association). The CVMA is developing non-binding guidelines on the welfare of animals (including farmed). Despite minimal legislation on animal welfare, the Chinese Government approves activities on animal welfare. A good example is the International Cooperation Committee on Animal Welfare (International Cooperation Committee on Animal Welfare (ICCAW), accessed on 4 March 2019), a non-profit organization approved by the Ministry of Agriculture and Rural Affairs that facilitates communication within the China livestock industry and international welfare organisations such as World Animal Protection and Compassion in World Farming.

The ICCAW (together with stakeholders) has published basic welfare standards for the main farmed species: pigs, chicken (laying hens, broilers), beef cattle and sheep [64]. The standards include guidance on enrichment (standard 5.7) referring to adequate manipulable materials (in line with scientific knowledge) to prevent abnormal, and allow, natural behaviour. These standards provide the first set of (voluntary) guidelines for pig welfare in China.

Independent of these activities, progressive businesses develop their own welfare programs. The first animal welfare issues to address are reducing the time pregnant sows spend in sow stalls and abolishing these altogether and move to group housing as well as to provide enrichment to all pigs. Collaborative projects have developed at an academic level [65] and on a more practical on-farm level [66].

Despite the existence of the ICCAW pig guidelines that mention enrichment, there are limitations as to what can be achieved on farms. Food safety is currently one of the main concerns of the Chinese pork consumer due to a string of food poisoning incidents in recent decades (see [67] for overview). This has had repercussions as consumers seem to prefer pork from industrialised farms as they associate that with constant quality and high food safety levels of the final product [ibid.].

This focus on food safety poses a main constraint on what materials farms are willing to use, as it severely restricts the use of substrates as enrichment materials, unless the farms know that these materials originate from safe sources. The consequence of this limitation is that when enrichment is provided, it often consists of a simple point-source object that does not necessarily meet all the requirements of effective enrichment (see Section 2.1).

Progress on farms can also be hampered due to a low awareness of animal welfare (both as a concept and on how this works in practice), and staff often have low technical skills, coupled with a high staff turnover. There is no effective management and infrastructure to collect information of on-farm practises at the government level [68], which will make encouraging, implementing and enforcing higher welfare practises very difficult. Another challenge for companies is that the marketing of products that are produced under higher welfare conditions is difficult with low consumer awareness of animal welfare. However, some consumers believe that when animals are treated well, their products taste better than regular products and this may provide marketing opportunities.

3.3. European Union

The EU is the second largest meat producing region worldwide [42]. Within the EU, pig meat represents 9% of agricultural output. While pig meat production has remained stable over the years, the total number of pigs has decreased since 2004. This reflects efficiency gains in pig farming (larger pig farms with more sows) despite stronger regulatory constraints (e.g., on welfare, such as group housing for sows, leading to increasing costs per sow) [69].

Compared to the USA and China, the European consumer appears more aware of farm animal welfare issues and also expresses clear views on this. In the most recent Eurobarometer survey [6], the majority of EU citizens interviewed (82%) believed that the welfare of farmed animals should be better protected than it is now, and 64% agreed with the statement that they would like to have more information about the conditions under which farmed animals are treated in their respective countries.

Legislation is the most common type of policy instrument used by the EU to achieve minimum welfare standards for farm animals. The aim of the policy is to spare animals all unnecessary suffering in three main phases of their lives: farming, transport and slaughter and to prohibit some of the most inhumane aspects of intensive, industrial livestock production [70]. The EU's approach to animal welfare legislation reflects changes in scientific understanding, changes in animal management practices, socio-economic studies and increasing consumer concern for animals (often for ethical reasons) and public expectations [71] and this has led to legislated animal welfare standards that are the most stringent in the world.

Pigs are protected under a specific directive (directives are a policy instrument in which the results to be achieved are binding, but each Member State's national authorities can choose the form and methods to get these results) that requires the provision of enrichment to all pigs. The relevant sections of the Pig Directive (2008/120/EC) that refer to enrichment are shown in Table 2.

Table 2. Sections of the EU Pig Directive that refer to enrichment materials [72].

Directive Section	Referring to:	Text
Article 3 (5.)	Sows and gilts	Member states shall ensure that, without prejudice to the requirements laid down in Annex I, sows and gilts have permanent access to manipulable material at least complying with the relevant requirements of that Annex.
Annex 1, Chapter 1 (4.)	All pigs	Notwithstanding Article 3 (5.), pigs must have permanent access to a sufficient quantity of material to enable proper investigation and manipulation activities, such as straw, hay, wood, sawdust, mushroom compost, peat or a mixture of such, which does not compromise the health of the animal.
Annex I, Chapter II: B. 3.	Sows and gilts	In the week before the expected farrowing time sows and gilts must be given suitable nesting material in sufficient quantity unless it is not technically feasible for the slurry system used in the establishment.
Annex I, Chapter II: C. 1.	Piglets	A part of the total floor, sufficient to allow the animals to rest together at the same time, must be solid or covered with a mat, or be littered with straw or any other suitable material.
Annex I, Chapter II: D. 3.	Weaners and rearing pigs	When signs of severe fighting appear, the causes shall be immediately investigated and appropriate measures taken, such as providing plentiful straw to the animals, if possible, or other materials for investigation. Animals at risk or particularly aggressive animals shall be kept separate from the group.

Legislation is not the only pathway to higher welfare. Private or voluntary animal welfare initiatives such as (welfare) farm assurance schemes (Notable examples are Soil Association organics and RSPCA Assured in the UK, Beter Leven in The Netherlands, Neuland in Germany, KRAV organics in Sweden and Label Rouge in France) promote good husbandry practice on farms and drive production standards towards higher welfare [73]. These schemes also increase compliance with animal welfare legislation as they are based on legal standards, coupled with regular audits [74,75]. Animal welfare assurance schemes (and associated food labelling) allow consumers to avoid the need to reflect on their food choice and the animals from which such products were derived, by delegating responsibility for issues such as animal welfare onto other actors, such as the state, supermarkets or brands [76]. Although in actual fact, consumers are often ignorant about the animal welfare standards within a particular scheme [77].

The European legal landscape, combined with animal welfare assurance schemes and individual business initiatives, provides a wealth of practical guidance for higher welfare in intensive pig production. There has been much recent activity to better implement the Pig Directive 2008/120/EC, with a focus on moving away from tail docking [78,79]. These studies show that, in countries that have banned tail docking (Finland, Sweden [80]), producers mostly provide straw and about 50% of the growing pigs in the UK are kept on straw [81]. Additionally, producers from several other countries are finding solutions to rear pigs with tails that involves providing effective enrichment. However, to be successful, this also requires a substantial change of management, with careful animal care, lower stocking densities, improved air and water quality, and the provision of a stress-free environment, which most farmers with existing systems find very difficult to achieve [78]. Furthermore, as outlined in Section 2, enrichment should not only be provided in systems that rear pigs with intact tails.

So, despite some success stories, the reality is much bleaker. Producers do not always provide enrichment and, if they do, it is mainly provided to intensively farmed weaned and growing/fattening pigs and mostly achieved with the provision of 'toys' (point-source enrichment-objects). The lack of providing material for manipulation to pigs constitutes a regular non-compliance with the Pig Directive [78]. These infringements have been highlighted following several inspection visits by the Food and Veterinary Office (now called Health, Food Audits and Analysis department of DG For Health and Food Safety) in several Member States (e.g., 82 FVO mission reports on animal welfare in the period 2005–2010 [82,83]. More recently, an EU Health and Food audit in Italy highlighted a lack of guidance on enrichment provision (against a background of managing risks to tail biting) [84].

The EU highlighted its intention on improving enforcement of legal requirements in its Strategy for the Protection and Welfare of Animals 2012–2015 [85].

Animal welfare NGOs have also focused on this issue, for example by publishing a compilation of photos taken from publications and websites from the EU pig industry [86], showing the absence of enrichment. While this is neither a very structured nor scientific approach, it does highlight how enrichment is regularly lacking as it should be visible in every pig pen, in the same way as feeders and drinkers are. When enrichment is provided, it is often an indestructible simple object which has no enduring value to the pig.

3.4. Common Deviations on EU Farms

3.4.1. Non-Effective or Hazardous Objects

Non-effective objects are very often provided. These objects do not meet the needs of pigs (as defined above) and can pose a hazard as they are unsafe (e.g., ingestible metal parts in car tyres [35]). Many studies have confirmed that objects such as chains, plastic pipes and balls or car tyres should not be recommended for long-term use, as they are not effective enrichment (e.g., [87,88]), do not meet the criteria of effectiveness and can quickly lose their novelty factor [1]. Composite objects are also often provided, e.g., by adding balls, pipes or pieces of hard wood to a metal chain. However, pig experts agree that trying to improve a metal chain in this way, only marginally improves welfare, remaining well below what they consider acceptable enrichment [81].

There is also concern that the manipulation of inadequate materials causes frustration when pigs try to interact with them (e.g., trying to bite in a ball or wooden log that is too wide for their mouths) [Van de Weerd, Pers. obs.]. Despite this, the use of such objects is still fairly widespread on farms, e.g., small balls on chains are prevalent in the Netherlands and Germany, while barren chains (without added destructible objects) appear to be prevalent in France and Belgium [81], and chains and plastic objects on farms in the UK [82]. Furthermore, the provision of materials such as solid wooden blocks, which are widely provided and perceived to be accepted as sufficient to comply with the EU Directive, can result in high levels of tail biting, especially if tail docking is not also used [89].

There are signs of progress, for example the debate in the Netherlands has moved on from being focused on implementing enrichment per se to the type of enrichment provided, with the government bodies recently announcing that they will enforce measures against inadequate enrichment such as a ball on a chain [90].

3.4.2. Inadequate Presentation

Other common mistakes are that (non-effective) objects are presented in the wrong way, such as loose on the floor (e.g., plastic canisters or balls) or hanging too high for pigs to reach (pigs are not able to lift their heads very high due to the anatomy of their neck). Although species-typical pig exploration is mainly directed at the ground level, suspending and attaching (with objects at eye or floor level) are favoured enrichment characteristics, because attachment prevents objects lodging in corners, becoming stuck behind feeders or getting pushed into neighbouring pens, out of reach [30,32,91].

3.4.3. Inadequate Location

Point-source objects as well as substrate racks or dispensers may be presented in the wrong location, with the only consideration being ease of maintenance (e.g., refill) for animal care takers. Objects suspended over the areas where most pigs sleep will lead to disruption of sleep patterns [1]. Enrichment should therefore be provided close to (but not in) the areas that are used for eating, drinking and elimination. This means that pigs who are active (move, eat, drink or manipulate enrichment) do not disturb sleeping pigs and enrichment does not get soiled with manure.

The location of substrate racks for group-housed (pregnant) sows needs careful consideration, especially in dynamic groups, as there will be high levels of activity near the racks, especially immediately post-feeding. Enrichment objects can be used to decrease activity around an electronic feeding station that normally attracts queuing sows, by leading animals away from the waiting and feeding station areas. Additionally, enrichment provided in a corner where only a small number of pigs can gain access can cause problems such as competition, aggression and displacements (see next section).

3.4.4. Inadequate Quantity/Size

If point-source enrichment is too small or of limited quantity, this will restrict availability [92], especially when grouped pigs synchronise their interactions with enrichment [36,93]. Limited accessibility to enrichment materials or objects may lead to social competition, aggression or restlessness and the redirection of exploration behaviour to pen structures or pen mates [93]. Distributing enrichments throughout a pen will help to reduce any negative effects of social status in sows [27] and reduce displacements [94]. Increasing object size will allow more pigs access simultaneously, but ease of access is also related to group size.

The size of enrichment objects should also be adjusted to the age and size of the pigs that they are presented to. If pigs are not able to adequately bite enrichment (grab between their jaws), they will be less able to chew, manipulate and possibly ingest, making the object less effective (see Table 1). This could explain why wooden logs with a diameter of 10 cm appeared less suitable as enrichment for weaner pigs (around 6.5 kg) compared with chains [95], and in comparison with wooden logs with a diameter of 10 cm provided to grower pigs of around 20 kg (and upwards) [23]. However, the method of attachment (in a wall-mounted bracket, or on loose chains), also differed between these two studies.

Systematic research into the number and placements of point-source objects in relation to group size would be valuable to elucidate these factors [1].

3.4.5. Inadequate Maintenance

Point-source objects are often put in a pen before a new batch of pigs arrive, but are subsequently not maintained, cleaned or renewed (if destroyed). Substrate bedding needs to be topped up daily to stay fresh and clean. It remains unclear whether pigs are deterred by soiled enrichment objects, an issue first raised by Blackshaw et al. [91]. Recent studies confirm lower interaction with logs presented on the floor, versus a hanging log, possibly caused by soiling with faeces [87]. However, the study on this issue by [96] found that cleaning (plastic and rubber) objects did not affect the pigs' interaction with them.

There are many aspects to consider with regards to optimal renewal or replacement rates, as long as the enrichment continues to provide meaningful reinforcement, it will retain enrichment value [17]. The risk of an object not retaining its enrichment value is illustrated by the pigs in a study by [87] who lost interest in non-effective enrichment objects (poplar wooden logs presented on chains or on the floor) quickly over a period of 6 weeks, whereas levels of tail and ear biting stayed at similar levels, thus increasing the risk of a tail-biting outbreak at some point.

Pigs can retain a memory of objects and therefore it is recommended to have a continuous rotation of point-source objects between pens, whereby the object exposure is limited to not more than two days and the same object is not introduced again until five days have passed, as this may help to preserve the exploratory value (novelty) of the objects rotated [97].

4. Main Barriers (Globally) and How to Overcome These

The pace and uptake of change in farm animal welfare is slow despite the demonstrable benefits of changes to the animals concerned. This suggests significant and persistent barriers to the uptake of practically applicable knowledge [98]. The overview of the situation in the main global pig-producing regions shows that this is clearly the case for enrichment for pigs. This barrier and other potential barriers, as well as possible ways to overcome these, will be discussed in this section.

4.1. Knowledge Transfer and Training

Despite the existing body of applied animal welfare knowledge on pig enrichment, there is still a need for more fundamental knowledge by applying theoretical frameworks. Tarou and Bashaw [17] describe several well-studied principles from the field of experimental analysis of behaviour and learning theory [99] as applied to enrichment for laboratory and zoo animals. Their approach is very useful for evaluating the short- and long-term effectiveness of enrichment and testing their predictions remains crucial for understanding and increasing the efficacy of enrichment programs for all captive animals, including farm animals.

There is an expectation that those who work with farm animals should know about animal welfare and animal behaviour issues, but there may still be a big gap in technical knowledge of farm staff, on pig-specific behaviours and on how motivated behaviour can channel into adverse behaviour, especially in restricted barren environments (e.g., the need to explore is channelled into tail biting [14]). Ultimately, influencing the behavioural attitude of farmers and animal caretakers can have a great effect on an animal's well-being and attitudes can change with new experiences or information about the animals [100].

Opportunities for welfare knowledge transfer include training and the use of advisory and extension services. These services give farmers and other food-chain actors access to information and knowledge, and, in return, inform research communities of the circumstances of its application in farm practice [98]. This ensures feedback between knowledge generation and its implementation. Furthermore, effective information delivery is dependent on understanding the motivations of animal care-takers and farmers as this is a pre-requisite for their potential receptivity of advice [101].

A fundamental requirement for the success of knowledge transfer will depend on the availability of good resources on enrichment [18,33] and training facilities [102,103] and suitably skilled people who can assist with implementing this knowledge in practice and present information accurately and without bias. Successful approaches to delivering animal welfare messages may have to utilise less unidirectional strategies, and more collaborative and participatory approaches. Such a method is successfully used in China by the NGO World Animal Protection [66]. Participatory approaches are effective in identifying and recognising common problems and creating common strategies for addressing them. In doing so, they allow farmers to retain ownership and control over possible solutions and methods to achieve them [98].

4.2. Economics (Return on Investment)

Applied animal welfare science does not have a great track record in linking animal welfare outcomes to economic benefits and the focus of enrichment programmes has often been on costs (see, e.g., the costs of providing wood enrichment in [104]). However, there has to be an obvious return on investment for producers to adopt effective enrichment strategies, otherwise it poses serious barriers for implementation [1] and hampers the ability or will of the pig industry to develop innovative enrichment.

When farms implement enrichment, their biggest investment is often not so much the cost of the materials, but investment in staff time to manage extra work (e.g., installing and maintaining enrichment, monitoring animals [89] and investment in staff (animal welfare) knowledge and technical skills. These issues are closely related to suggested barriers to abandon tail docking [78]. The focus for progress has to be on the returns that high standards of animal care can yield. The costs and associated economic benefits of providing enrichment and reduce damaging behaviours such as ear and tail biting have so far been mostly studied.

The costs of tail biting have been assessed and range from an extra net cost of €18.96 for a victim of tail biting, impacting on the net margin of production (modelled data) [89], and up to 43% less profit per pig, due to carcass condemnation and trimming losses (slaughter house study) [105]. Furthermore, daily weight gain in tail-bitten growing pigs can be reduced by 1 to 3% [106].

Disease is a concomitant threat to animal welfare and business sustainability, but few studies have focused on the relationship between enriched environments and health issues or disease susceptibility. Some research shows the potential of studying this relationship. For example, straw bedding reduced the incidence of gastric lesions in growing pigs, reflecting either lower levels of stress as compared to barren housing or a positive effect of stomach content firmness [107]. Enriched housed pigs showed less stress-related behaviour and had reduced disease susceptibility to co-infection of PRRSV (Porcine Reproductive and Respiratory Virus) and *A. pleuropneumoniae* (associated with fewer lesions in the lungs, and a lower total pathologic tissue damage score) [108].

For group-housed pregnant sows there is even less scientific information on the relationship between the facilities in a group-housing system and infectious disease, although the fact that the lying and defecation areas are separated improves hygiene and decreases the intensity of oral contact with faeces [109].

There are also possible gains in terms of performance that need to be quantified, e.g., crate-born piglets raised with enrichment from birth can have improved growth and a modulated immune response compared to piglets from barren crates [110]. Future efforts should focus on quantifying the benefits of enrichment in relation to the improved performance of growing pigs [32], improved performance of sows (e.g., lower lameness levels, lower stress levels and stereotypies, improved pregnancies, ease of farrowing, shorter partus, increased maternal care), possibly leading to more pigs/sow/year; and possible improved piglet survival (e.g., coping with and adapting to weaning, reducing the growth check after weaning, e.g., [111]).

There is a clear role for applied ethologists together with agriculture economists to describe these economic benefits of higher welfare strategies (or adversely, the costs of low welfare). Mentioning economics raises the issue of who should pay for higher welfare production with the finger often pointing at consumers who should be willing to pay the increased costs [89] via a price premium for higher welfare products (see, e.g., 'Heart Pig' higher animal welfare brand (Denmark), with sows mainly in loose housing, growing pigs with 10% more space and not tail docked and with continuous access to straw in racks. The production costs are 7.9% higher (e.g., covering increased costs for more, straw, mortality, labour), rising from €1.41/kg up to €1.52/kg slaughter weight. These higher costs are covered by a price premium for the meat, approximately €0.17 per kilo [112]. However, we may have to think broader and re-invent the business model for higher welfare farming and re-think how welfare is valued in the supply chain, for example by applying similar methods as those applied to environmental monitoring services [113]. The participative process presented by these authors combines open innovation in idea generation, evaluation and the development of ideas and identification of a new core business model. Such a model should build on adequate compensation and incentives for producers that collect and contribute valuable animal welfare data.

4.3. Novel Drivers of Change

The commitments of food companies can be a major driving force to influence the welfare of animals, especially if the commitments expressed are translated into actual behaviour [11]. There is increasing pressure on food companies to include animal welfare as an area of focus within their business objectives, especially those that have animal products in their supply chain (e.g., food retailers, processors and food service businesses).

Corporate Social Responsibility (CSR) is the principle instrument to communicate a company's ethical and social commitments [114] and CSR statements are also an important component of branding, assuring consumers that due diligence requirements are being met and helping to retain customer fidelity [98]. Many companies have incorporated animal welfare statements within their social responsibility targets [11,115]. Furthermore, companies are starting to realise that ignoring supply chain-related animal welfare issues may create business and brand value risks [114,116], and this increases pressure on companies to include animal welfare as a focus of management. However, despite this pressure, the topic has remained a largely immature issue [117].

Until recently, investors regarded farm animal welfare as a niche ethical issue rather than a business issue and assumed that higher welfare farming inevitably results in higher financial costs for companies. Investors are an important influence on how companies manage the social and environmental impacts of their operations, and as such they can exert pressure on the behaviour of companies via their role as stakeholders or via the views they express when meeting with companies [118]. Some investor initiatives highlight the risks of intensive farming beyond animal welfare per se. These risks include threats to food safety, nutrition and public health (including antimicrobial resistance, the environment and labour rights, e.g., [119]).

Tools such as the Business Benchmark on Farm Animal Welfare (BBFAW, an annual evaluation of the management of farm animal welfare in the world's largest food companies) have helped investors to differentiate between those companies that manage farm animal welfare well and those that do not, as the BBFAW encourages improved reporting on farm animal welfare by companies and it provides a robust framework for assessing companies' approaches to this issue [118]. There is evidence that companies respond to signals sent to them via responsible investors, as there has been a year on year increase in the overall BBFAW score for most of the businesses assessed, despite tightening of the Benchmark criteria and the increased emphasis on performance reporting and impact [11,120].

However, there is no room for complacency as about 47% of companies benchmarked in 2018 provided little or no information on their approach to farm animal welfare and this suggests that there is more to do, both in terms of encouraging improvements in policies, management systems and processes, and in ensuring that improvements are institutionalised and maintained over time [120].

To date, the BBFAW has not specifically assessed companies on policies requiring enrichment for pigs, but it does assess companies' policies on routine mutilations. For pigs, these include tail docking, castration and teeth clipping. The proportion of companies that published policies in this area broadly increased over the benchmarked years (2012–2018). However, in 2018, only 24% of companies (with pigs in their supply) had made current (or future) public commitments specifically prohibiting the routine use of tail docking in pigs (data from [120] analysed). Furthermore, none of the companies accompanied commitments on tail docking with requirements for the provision of enrichment which is imperative if tail docking is not performed.

5. Conclusions

In the coming decades, the sustainability of pig production systems will not only rely on efficiency improvements at the herd level, but also on other factors: an increased use of alternative feed sources; reduced crude protein content in the rations; the proper use of pig manure as fertilizer through crop-livestock reconnection; the moderation of the human demand for pork; reducing the use of antibiotics [41,121]. We should add to that list the humane treatment of pigs and a focus on higher welfare production systems. This will involve extending the fundamental scientific basis for enrichment and building on what we already know and then keep channelling that knowledge to farms via advice and training.

With regards to regional progress, China is very much at the beginning of its pig welfare journey, against a background that the understanding of welfare in the general population is not particularly advanced. The USA has just passed the initial stages and is moving towards applying and generating animal welfare science and putting that into practice. The challenge for the US pig industry is to utilise the existing knowledge and extend it. The EU is well on its way on its welfare journey. However, the examples of issues on EU farms illustrate the challenge that the EC Pig Directive poses, especially with regards to the permanent provision of destructible materials. This requires continuous inspection and enforcement of the legislation. Table 3 summarises the current situation and likely future progress in the top pig-producing regions.

Table 3. Summary of the progress and likely future drivers for change in the journey towards implementing effective environmental enrichment for pigs in the World's largest pig-producing regions (China, the European Union (EU) and the USA cover 77% of global production combined).

Region	Driver for Change			
	Regulation	Consumer/NGO Pressure	Guidelines/Assurance Schemes	Food Business/CSR Driven
USA	• Some state restrictions on sow stalls and outright cruelty, no other federal or state legislation covering pig welfare during rearing. • Unlikely to be a significant driver for change at the federal level but some progress may be made at the state level.	• NGO pressure has focused on sow confinement with some progress and likely to extend to other pig welfare issues including enrichment. • More information on consumer perceptions of unenriched environments and willingness to pay is needed.	• Industry-led schemes exclude enrichment provision. • A few third-party auditing and labelling schemes exist and include enrichment provision. • Consumer demand for niche products is a relatively small but increasing market share.	• Changing food business policy has been a significant driver for change in the case of sow stalls. • In some cases, food business policy includes enrichment provision. • Likely to be the biggest driver for change at scale in the future.
China	• Legislation is minimal and mainly focuses on food safety rather than animal protection. • Could be a driver for change in the future as the government approves of animal welfare activities.	• Consumer awareness of animal welfare is low but increasing, particularly in large cities. • NGOs are working on pig welfare and on increasing consumer awareness of pig welfare issues.	• The CVMA [a] is developing non-binding animal welfare guidelines. • ICCAW [b] pig welfare guidelines include enrichment. • Third-party auditing and labelling schemes likely to appear in the future.	• A few progressive businesses have their own pig welfare policies. • Difficult to market products due to low consumer awareness. • Likely to be a future driver with increasing awareness and interest in the BBFAW [c].
EU	• Pigs are protected under a Directive that requires the provision of enrichment to all pigs. • However, lack of adequate enrichment is a regular non-compliance to the Directive.	• NGOs work both at the EU and country level to support the implementation of, or go beyond, minimum standards. • European consumers appear more aware of farm animal welfare and the majority (82%) of those asked believed it should be better protected.	• Private or voluntary initiatives promote pig welfare. • These schemes increase compliance with legislation and drive welfare standards higher. • Likely to increase in popularity and extend to a greater number of member states.	• Many food businesses have progressive pig welfare policies to meet consumer expectations. • Businesses, particularly supermarkets, can gain a competitive advantage. • Financial institutions now also exerting pressure on food businesses to raise standards.

[a] Chinese Veterinary Medical Association; [b] International Cooperation Committee on Animal Welfare; [c] Business Benchmark for Farm Animal Welfare.

Designing effective enrichment is not an easy task, considering the range of aspects to consider in terms of design and presentation and the labour involved in implementing such a programme. Even when the guidance on effective enrichment has been followed (Section 2.1), the enrichment has to be assessed in situ to assess whether it is having the intended effect (e.g., by regular observations of the animals' behaviour, such as aggression, displacements and vocalizations). This assessment should not only focus on short-term effects but monitor the long-term benefits of the enrichment strategy used, with a continuous cycle of evaluation and improvement. Producers may benefit from tailored advice that is specific to their situation [122] and the specific issues that they face on their farm and from the market they operate in.

There is increasing pressure from the financial sector (and some of the published guidance includes the provision of enrichment for pigs [123] on food companies to include animal welfare as an area of focus within their business objectives, and this can lead to tangible change for animals on farms. Highlighting the economic benefits of effective enrichment and focusing on the returns of investment will help to break down barriers to further investment.

The pig production community has had more than 50 years to implement and combine scientific and practical knowledge but is still a long way off reaching the ultimate destination of an enriched pig population.

Author Contributions: Conceptualization, H.v.d.W.; Investigation, H.v.d.W.; Writing—Original draft, H.v.d.W.; Writing—review and editing, H.v.d.W. and S.I.

Funding: This research received no external funding.

Acknowledgments: We are grateful to the World Animal Protection China team for supplying information on their work with producers to improve pig welfare. We thank Jon Day for helpful comments on the draft manuscript.

Conflicts of Interest: The authors declare no conflict of interest.

References

1. van de Weerd, H.A.; Day, J.E.L. A review of environmental enrichment for pigs housed in intensive housing systems. *Appl. Anim. Behav. Sci.* **2009**, *116*, 1–20. [CrossRef]
2. Newberry, R.C. Environmental enrichment: Increasing the biological relevance of captive environments. *Appl. Anim. Behav. Sci.* **1995**, *44*, 229–243. [CrossRef]
3. Stolba, A.; Wood-Gush, D.G. The identification of behavioural key features and their incorporation into a housing design for pigs. *Ann. Rech. Vet.* **1984**, *15*, 287–299. [PubMed]
4. Te Velde, H.; Aarts, N.; Van Woerkum, C. Dealing with ambivalence: Farmers' and consumers' perceptions of animal welfare in livestock breeding. *J. Agric. Environ. Ethics* **2002**, *15*, 203–219. [CrossRef]
5. Spooner, J.M.; Schuppli, C.A.; Fraser, D. Attitudes of Canadian citizens toward farm animal welfare: A qualitative study. *Livest. Sci.* **2014**, *163*, 150–158. [CrossRef]
6. European Commission. Special Eurobarometer 442 Report Attitudes of Europeans towards Animal Welfare. Available online: http://www.eurogroupforanimals.org/eurobarometer (accessed on 15 February 2019).
7. Beltran-Alcrudo, D.; Falco, J.R.; Raizman, E.; Dietze, K. Transboundary spread of pig diseases: The role of international trade and travel. *BMC Vet. Res.* **2019**, *15*, 64. [CrossRef] [PubMed]
8. Alcorn, T.; Ouyang, Y. China's invisible burden of foodborne illness. *Lancet* **2012**, *379*, 789–790. [CrossRef]
9. Mellor, D.J.; Webster, J.R. Development of animal welfare understanding drives change in minimum welfare standards. *Rev. Sci. Tech. Off. Int. Epiz* **2014**, *33*, 121–130. [CrossRef]
10. Fraser, D. The globalisation of farm animal welfare. *Rev. Sci. Tech. Off. Int. Epiz* **2014**, *33*, 33–38. [CrossRef]
11. Sullivan, R.; Amos, N.; van de Weerd, H.A. Corporate reporting on farm animal welfare: An evaluation of global food companies' discourse and disclosures on farm animal welfare. *Animals* **2017**, *7*, 17. [CrossRef]
12. Millman, S.T.; Duncan, I.J.H.; Stauffacher, M.; Stookey, J.M. The impact of applied ethologists and the International Society for Applied Ethology in improving animal welfare. *Appl. Anim. Behav. Sci.* **2004**, *86*, 299–311. [CrossRef]
13. Young, R. *Environmental Enrichment for Captive Animals*; Blackwell Publishing: Oxford UK, 2003; ISBN 0-632-06407-2.

14. Fraser, D. The role of behavior in swine production: A review of research. *Appl. Anim. Ethol.* **1984**, *11*, 317–339. [CrossRef]
15. Van De Weerd, H.A.; Docking, C.M.; Day, J.E.L.; Edwards, S.A. The development of harmful social behaviour in pigs with intact tails and different enrichment backgrounds in two housing systems. *Anim. Sci.* **2005**, *80*, 289–298. [CrossRef]
16. Zonderland, J.J.; van Riel, J.W.; Bracke, M.B.M.; Kemp, B.; den Hartog, L.A.; Spoolder, H.A.M. Tail posture predicts tail damage among weaned piglets. *Appl. Anim. Behav. Sci.* **2009**, *121*, 165–170. [CrossRef]
17. Tarou, L.R.; Bashaw, M.J. Maximizing the effectiveness of environmental enrichment: Suggestions from the experimental analysis of behavior. *Appl. Anim. Behav. Sci.* **2007**, *102*, 189–204. [CrossRef]
18. European Commission Staff Working Document on Best Practices with a View to the Prevention of Routine Tail-Docking and the Provision of Enrichment Materials to Pigs. Available online: https://ec.europa.eu/food/sites/food/files/animals/docs/aw_practice_farm_pigs_stfwrkdoc_en.pdf (accessed on 15 February 2019).
19. Taylor, N.R.; Main, D.C.J.; Mendl, M.; Edwards, S.A. Tail-biting: A new perspective. *Vet. J.* **2010**, *186*, 137–147. [CrossRef]
20. Brunberg, E.I.; Bas Rodenburg, T.; Rydhmer, L.; Kjaer, J.B.; Jensen, P.; Keeling, L.J. Omnivores going astray: A review and new synthesis of abnormal behavior in pigs and laying hens. *Front. Vet. Sci.* **2016**, *3*, 57. [CrossRef] [PubMed]
21. Wischner, D.; Kemper, N.; Krieter, J. Nest-building behaviour in sows and consequences for pig husbandry. *Livest. Sci.* **2009**, *124*, 1–8. [CrossRef]
22. Vanheukelom, V.; Driessen, B.; Geers, R. The effects of environmental enrichment on the behaviour of suckling piglets and lactating sows: A review. *Livest. Sci.* **2012**, *143*, 116–131. [CrossRef]
23. Telkänranta, H.; Swan, K.; Hirvonen, H.; Valros, A. Chewable materials before weaning reduce tail biting in growing pigs. *Appl. Anim. Behav. Sci.* **2014**, *157*, 14–22. [CrossRef]
24. Verdon, M.; Hansen, C.F.; Rault, J.; Jongman, E.; Hansen, L.U.; Plush, K.; Hemsworth, P.H. Effects of group housing on sow welfare: A review. *J. Anim. Sci.* **2015**, *93*, 1999–2017. [CrossRef] [PubMed]
25. Stewart, C.L.; O'Connell, N.E.; Boyle, L. Influence of access to straw provided in racks on the welfare of sows in large dynamic groups. *Appl. Anim. Behav. Sci.* **2008**, *112*, 235–247. [CrossRef]
26. Horback, K.M.; Pierdon, M.K.; Parsons, T.D. Behavioral preference for different enrichment objects in a commercial sow herd. *Appl. Anim. Behav. Sci.* **2016**, *184*, 7–15. [CrossRef]
27. Brown, J.A.; Roy, C.R.; Seddon, Y.M.; Connor, L.M. Effects of enrichment and social status on enrichment use, aggression and stress response of sows housed in ESF pens. In Proceedings of the 52nd Congress of the International Society for Applied Ethology, Charlottetown, PE, Canada, 30 July–3 August 2018; p. 231.
28. Greenwood, E.C.; van Wettere, W.H.E.J.; Rayner, J.; Hughes, P.E.; Plush, K.L. Provision point-source materials stimulates play in sows but does not affect aggression at regrouping. *Animals* **2019**, *9*, 8. [CrossRef]
29. Petak, I.; Mrljak, V.; Tadić, Z.; Krsnik, B. Preliminary study of breeding boars' welfare. *Vet. Arh.* **2010**, *80*, 235–246.
30. Van De Weerd, H.A.; Docking, C.M.; Day, J.E.L.; Avery, P.J.; Edwards, S.A. A systematic approach towards developing environmental enrichment for pigs. *Appl. Anim. Behav. Sci.* **2003**, *84*, 101–118. [CrossRef]
31. Bracke, M.B.M.; Zonderland, J.J.; Lenskens, P.; Schouten, W.G.P.; Vermeer, H.; Spoolder, H.A.M.; Hendriks, H.J.M.; Hopster, H. Formalised review of environmental enrichment for pigs in relation to political decision making. *Appl. Anim. Behav. Sci.* **2006**, *98*, 165–182. [CrossRef]
32. Averós, X.; Brossard, L.; Dourmad, J.Y.; de Greef, K.H.; Edge, H.L.; Edwards, S.A.; Meunier-Salaün, M.C. A meta-analysis of the combined effect of housing and environmental enrichment characteristics on the behaviour and performance of pigs. *Appl. Anim. Behav. Sci.* **2010**, *127*, 73–85. [CrossRef]
33. European Commission. Cutting the Need for Tail Docking. Available online: https://ec.europa.eu/food/animals/welfare/practice/farm/pigs/tail-docking_en (accessed on 15 February 2019).
34. van de Weerd, H.A. Appropriate Enrichment. In *Animal Welfare in Practice: Pigs*; Camerlink, I., Ed.; 5M Publishing: London, UK, 2019; In press.
35. EFSA Scientific Opinion concerning a Multifactorial approach on the use of animal and non-animal-based measures to assess the welfare of pigs. *Eur. Food Saf. Auth. J.* **2014**, *12*, 1–101.
36. Bulens, A.; Van Beirendonck, S.; Van Thielen, J.; Buys, N.; Driessen, B. Straw applications in growing pigs: Effects on behavior, straw use and growth. *Appl. Anim. Behav. Sci.* **2015**, *169*, 26–32. [CrossRef]

37. Telkänranta, H.; Bracke, M.B.M.; Valros, A. Fresh wood reduces tail and ear biting and increases exploratory behaviour in finishing pigs. *Appl. Anim. Behav. Sci.* **2014**, *161*, 51–59. [CrossRef]
38. Trickett, S.L.; Guy, J.H.; Edwards, S.A. The role of novelty in environmental enrichment for the weaned pig. *Appl. Anim. Behav. Sci.* **2009**, *116*, 45–51. [CrossRef]
39. Day, J.E.L.; Kyriazakis, I.; Rogers, P.J. Food choice and intake: Towards a unifying framework of learning and feeding motivation. *Nutr. Res. Rev.* **1998**, *11*, 25–43. [CrossRef]
40. van Dijk, L.; Buller, J.H.; Blokhuis, J.H.; van Niekerk, T.; Voslarova, E.; Manteca, X.; Weeks, A.C.; Main, C.D. HENNOVATION: Learnings from Promoting Practice-Led Multi-Actor Innovation Networks to Address Complex Animal Welfare Challenges within the Laying Hen Industry. *Animals* **2019**, *9*, 24. [CrossRef] [PubMed]
41. Lassaletta, L.; Estellés, F.; Beusen, A.H.W.; Bouwman, L.; Calvet, S.; van Grinsven, H.J.M.; Doelman, J.C.; Stehfest, E.; Uwizeye, A.; Westhoek, H. Future global pig production systems according to the Shared Socioeconomic Pathways. *Sci. Total Environ.* **2019**, *665*, 739–751. [CrossRef]
42. Agri Benchmark. *Pig Report. Understanding Agriculture Worldwide*. 2018. Available online: http://catalog.agribenchmark.org/blaetterkatalog/Pig_Report_2018/#page_1 (accessed on 6 March 2019).
43. Sexton, A.E.; Garnett, T.; Lorimer, J. Framing the future of food: The contested promises of alternative proteins. *Environ. Plan. E Nat. Space* **2019**, *2*, 47–72. [CrossRef]
44. World Animal Protection. Country Report USA 2014. Available online: https://api.worldanimalprotection.org/country/usa (accessed on 4 March 2019).
45. The Humane Society of the United States. *Humane State Ranking 2018 (Alabama through Missouri)*. Available online: https://www.humanesociety.org/search?keys=humane+state+ranking (accessed on 28 March 2019).
46. The Humane Society of the United States. *Humane State Ranking 2018 (Montana through Wyoming)*. Available online: https://www.humanesociety.org/search?keys=humane+state+ranking (accessed on 28 March 2019).
47. The Humane Society of the United States. *Humane State Ranking 2018: Total Scores*. Available online: https://www.humanesociety.org/search?keys=humane+state+ranking (accessed on 28 March 2019).
48. Picardy, J.A.; Pietrosemoli, S.; Griffin, T.S.; Peters, C.J. Niche pork: Comparing pig performance and understanding producer benefits, barriers and labeling interest. *Renew. Agric. Food Syst.* **2017**, *34*, 7–19. [CrossRef]
49. Honeyman, M.S. Extensive bedded indoor and outdoor pig production systems in USA: current trends and effects on animal care and product quality. *Livest. Prod. Sci.* **2005**, *94*, 15–24. [CrossRef]
50. National Pork Board. *PQA Plus Education Handbook*; National Pork Board: Des Moines, Iowa, USA, 2016; pp. 1–128.
51. National Pork Board. *PQA Plus Site Assessment Guide*; National Pork Board: Des Moines, Iowa, USA, 2016; pp. 1–68.
52. Apple, J.K.; Craig, J.V. The influence of pen size on toy preference of growing pigs. *Appl. Anim. Behav. Sci.* **1992**, *35*, 149–155. [CrossRef]
53. Ison, S.H.; Bates, R.O.; Ernst, C.W.; Steibel, J.P.; Siegford, J.M. Housing, ease of handling and minimising inter-pig aggression at mixing for nursery to finishing pigs as reported in a survey of North American pork producers. *Appl. Anim. Behav. Sci.* **2018**, *205*, 159–166. [CrossRef]
54. Waters, J. Ethics and the choice of animal advocacy campaigns. *Ecol. Econ.* **2015**, *119*, 107–117. [CrossRef]
55. Shields, S.; Shapiro, P.; Rowan, A. A decade of progress toward ending the intensive confinement of farm animals in the united states. *Animals* **2017**, *7*, 40. [CrossRef]
56. The Cheesecake Factory Animal Welfare Update 2018. Available online: https://www.thecheesecakefactory.com/assets/pdf/The_Cheesecake_Factory_Animal_Welfare_Update_July_2018.pdf (accessed on 31 March 2019).
57. Mench, J.A. Farm animal welfare in the U.S.A.: Farming practices, research, education, regulation, and assurance programs. *Appl. Anim. Behav. Sci.* **2008**, *113*, 298–312. [CrossRef]
58. Hansen, J.; Gale, F. China in the Next Decade: Rising Meat Demand and Growing Imports of Feed. USDA Amber Waves, 1A. Available online: https://ageconsearch.umn.edu/record/211199/files/http---www_ers_usda_gov-amber-waves-2014-april-china-in-the-next-decade-rising-meat-demand-and-growing-imports-of-feed_aspx__Vi5BRMFIGIZ_pdfmyurl.pdf (accessed on 2 March 2019).
59. Qian, Y.; Song, K.; Hu, T.; Ying, T. Environmental status of livestock and poultry sectors in China under current transformation stage. *Sci. Total Environ.* **2018**, *622–623*, 702–709. [CrossRef] [PubMed]

60. You, X.; Li, Y.; Zhang, M.; Yan, H.; Zhao, R. A Survey of Chinese Citizens' Perceptions on Farm Animal Welfare. *PLoS ONE* **2014**, *9*, e109177. [CrossRef] [PubMed]
61. Sinclair, M.; Zito, S.; Phillips, C.; Sinclair, M.; Zito, S.; Phillips, C.J.C. The Impact of Stakeholders' Roles within the Livestock Industry on Their Attitudes to Livestock Welfare in Southeast and East Asia. *Animals* **2017**, *7*, 6. [CrossRef]
62. Lai, J.; Wang, H.H.; Ortega, D.L.; Olynk Widmar, N.J. Factoring Chinese consumers' risk perceptions into their willingness to pay for pork safety, environmental stewardship, and animal welfare. *Food Control* **2018**, *85*, 423–431. [CrossRef]
63. World Animal Protection. Country Report China 2014. Available online: https://api.worldanimalprotection.org/country/china (accessed on 4 March 2019).
64. International Cooperation Committee on Animal Welfare (ICCAW). Farm Animal Welfare Requirements Pigs. Available online: http://www.iccaw.org.cn/plus/list.php?tid=89 (accessed on 5 March 2019).
65. Zhou, Q.; Sun, Q.; Wang, G.; Zhou, B.; Lu, M.; Marchant-Forde, J.N.; Yang, X.; Zhao, R. Group housing during gestation affects the behaviour of sows and the physiological indices of offspring at weaning. *Animal* **2014**, *8*, 1162–1169. [CrossRef]
66. Ison, S.; Blaszak, K.; Mora, R.; Van de Weerd, H.A.; Kavanagh, L. Group sow housing with enrichment: Insights from Brazil, China and Thailand. In Proceedings of the 2019 Banff Pork Seminar, Banff, AL, Canada, 8–10 January 2019; Volume 30, 236p.
67. Cicia, G.; Caracciolo, F.; Cembalo, L.; Del Giudice, T.; Grunert, K.G.; Krystallis, A.; Lombardi, P.; Zhou, Y. Food safety concerns in urban China: Consumer preferences for pig process attributes. *Food Control* **2016**, *60*, 166–173. [CrossRef]
68. Wei, X.; Lin, W.; Hennessy, D.A. Biosecurity and disease management in China's animal agriculture sector. *Food Policy* **2015**, *54*, 52–64. [CrossRef]
69. Meat production statistics. Eurostat Statistics Explained. Available online: https://ec.europa.eu/eurostat/statistics-explained/index.php/Meat_production_statistics#Pigmeat (accessed on 11 March 2019).
70. Van de Weerd, H.A.; Day, J.E.L. Farm animal welfare: The legal journey to improved farm animal welfare. In *The Business of Farm Animal Welfare*; Amos, N., Sullivan, R., Eds.; Routledge: Abingdon, UK, 2018; pp. 47–63, ISBN 978-1-78353-529-3.
71. Horgan, R.; Gavinelli, A. The expanding role of animal welfare within EU legislation and beyond. *Livest. Sci.* **2006**, *103*, 303–307. [CrossRef]
72. EU Council. Council Directive 2008/120/EC of 18 December 2008 laying down minimum standards for the protection of pigs (codified version, consolidating earlier Directives adopted in 1991 and 2001). *Off. J. Eur. Union* **2009**, *L47*, 5–13.
73. Kilchsperger, R.; Schmid, O.; Hecht, J. *Animal Welfare Initiatives in Europe. Technical Report on Grouping Method for Animal Welfare Standards and Initiatives (EconWelfare Project Report D1.1.)*. Available online: https://cordis.europa.eu/project/rcn/87806/reporting/en (accessed on 12 March 2019).
74. KilBride, A.L.; Mason, S.A.; Honeyman, P.C.; Pritchard, D.G.; Hepple, S.; Green, L.E. Associations between membership of farm assurance and organic certification schemes and compliance with animal welfare legislation. *Vet. Rec.* **2012**, *170*, 152. [CrossRef]
75. Van Wagenberg, C.P.A.; Brouwer, F.M.; Hoste, R.; Rau, M.L. Comparative Analysis of EU Standards in Food Safety, Environment, Animal Welfare and Other Non-Trade Concerns with Some Selected Countries. Available online: http://www.europarl.europa.eu/RegData/etudes/etudes/join/2012/474542/IPOL-AGRI_ET.pdf (accessed on 10 March 2019).
76. Evans, A.; Miele, M. *Consumers' Views about Farm Animal Welfare. Part II: European Comparative Report Based on Focus Group Research*. 2008. Available online: http://www.welfarequality.net/en-us/reports/ (accessed on 6 March 2019).
77. Duffy, R.; Fearne, A. Value perceptions of farm assurance in the red meat supply chain. *Br. Food J.* **2009**, *111*, 669–685. [CrossRef]
78. Nalon, E.; De Briyne, N. Efforts to Ban the Routine Tail Docking of Pigs and to Give Pigs Enrichment Materials via EU Law: Where do We Stand a Quarter of a Century on? *Animals* **2019**, *9*, 132. [CrossRef]
79. European Commission. Directorate Health and Food Audits and Analysis. Overview Report Study Visits on Rearing Pigs with Intact Tails. Available online: http://ec.europa.eu/food/auditsanalysis/overview_reports/act_getPDF.cfm?PDF_ID=790 (accessed on 6 March 2019).

80. Wallgren, T.; Westin, R.; Gunnarsson, S. A survey of straw use and tail biting in Swedish pig farms rearing undocked pigs. *Acta Vet. Scand.* **2016**, *58*, 84. [CrossRef]
81. Bracke, M.B.M.; Koene, P. Expert opinion on metal chains and other indestructible objects as proper enrichment for intensively-farmed pigs. *PLoS ONE* **2019**, *14*, e0212610. [CrossRef]
82. Pandolfi, F.; Stoddart, K.; Wainwright, N.; Kyriazakis, I.; Edwards, S.A. The "Real Welfare" scheme: Benchmarking welfare outcomes for commercially farmed pigs. *Animal* **2017**, *11*, 1816–1824. [CrossRef]
83. Rayment, M.; Asthana, P.; Van de Weerd, H.A.; Gittins, J.; Talling, J. *Evaluation of the EU Policy on Animal Welfare and Possible Options for the Future*. 2010. Available online: https://ec.europa.eu/food/sites/food/files/animals/docs/aw_arch_122010_full_ev_report_en.pdf (accessed on 12 March 2019).
84. FVO Animal Welfare-Tail-Docking of Pigs. FVO Report 2017-6257. Available online: http://ec.europa.eu/food/audits-analysis/audit_reports/details.cfm?rep_id=3987 (accessed on 12 March 2019).
85. EU Strategy for the Protection and Welfare of Animals 2012–2015. Available online: https://ec.europa.eu/food/animals/welfare/strategy_en (accessed on 30 March 2019).
86. Compassion in World Farming. Lack of Complicate with the Pigs Directive Continues: Urgent Need for Change. Available online: https://www.ciwf.org.uk/research/species-pigs/lack-of-compliance-with-the-pigs-directive-continues-urgent-need-for-change/ (accessed on 12 March 2019).
87. Giuliotti, L.; Benvenuti, M.N.; Giannarelli, A.; Mariti, C.; Gazzano, A. Effect of Different Environment Enrichments on Behaviour and Social Interactions in Growing Pigs. *Animals* **2019**, *9*, 101. [CrossRef]
88. Ishiwata, T.; Uetake, K.; Tanaka, T. Factors affecting agonistic interactions of weanling pigs after grouping in pens with a tire. *Anim. Sci. J.* **2004**, *75*, 71–78. [CrossRef]
89. D'Eath, R.B.; Niemi, J.K.; Vosough Ahmadi, B.; Rutherford, K.M.D.; Ison, S.H.; Turner, S.P.; Anker, H.T.; Jensen, T.; Busch, M.E.; Jensen, K.K.; et al. Why are most EU pigs tail docked? Economic and ethical analysis of four pig housing and management scenarios in the light of EU legislation and animal welfare outcomes. *Animal* **2016**, *10*, 687–699. [CrossRef]
90. Van der Plas, C. NVWA Gaat Strenger Controleren en Handhaven op Spelmateriaal voor Varkens. Available online: https://www.pigbusiness.nl/artikel/187360-nvwa-gaat-strenger-controleren-en-handhaven-op-spelmateriaal-voor-varkens/ (accessed on 6 March 2019).
91. Blackshaw, J.K.; Thomas, F.J.; Lee, J.-A. The effect of a fixed or free toy on the growth rate and aggressive behaviour of weaned pigs and the influence of hierarchy on initial investigation of the toys. *Appl. Anim. Behav. Sci.* **1997**, *53*, 203–212. [CrossRef]
92. van de Weerd, H.A.; Docking, C.M.; Day, J.E.L.; Breuer, K.; Edwards, S.A. Effects of species-relevant environmental enrichment on the behaviour and productivity of finishing pigs. *Appl. Anim. Behav. Sci.* **2006**, *99*, 230–247. [CrossRef]
93. Zwicker, B.; Weber, R.; Wechsler, B.; Gygax, L. Degree of synchrony based on individual observations underlines the importance of concurrent access to enrichment materials in finishing pigs. *Appl. Anim. Behav. Sci.* **2015**, *172*, 26–32. [CrossRef]
94. Zwicker, B.; Gygax, L.; Wechsler, B.; Weber, R. Influence of the accessibility of straw in racks on exploratory behaviour in finishing pigs. *Livest. Sci.* **2012**, *148*, 67–73. [CrossRef]
95. Nannoni, E.; Sardi, L.; Vitali, M.; Trevisi, E.; Ferrari, A.; Barone, F.; Bacci, M.L.; Barbieri, S.; Martelli, G. Effects of different enrichment devices on some welfare indicators of post-weaned undocked piglets. *Appl. Anim. Behav. Sci.* **2016**, *184*, 25–34. [CrossRef]
96. Beaudoin, J.M.; Bergeron, R.; Devillers, N.; Laforest, J. Growing Pigs' Interest in Enrichment Objects with Different Characteristics and Cleanliness. *Animals* **2019**, *9*, 85. [CrossRef]
97. Gifford, A.K.; Cloutier, S.; Newberry, R.C. Objects as enrichment: Effects of object exposure time and delay interval on object recognition memory of the domestic pig. *Appl. Anim. Behav. Sci.* **2007**, *107*, 206–217. [CrossRef]
98. Farm Animal Welfare Committee (FAWC). Report on Education about Farm Animal Welfare. Available online: https://www.gov.uk/government/publications/fawc-report-on-education-about-farm-animal-welfare (accessed on 4 March 2019).
99. Lieberman, D.A. *Learning: Behavior and Cognition*, 2nd ed.; Brooks/Cole Publishing Company: Pacific Grove, CA, USA, 1993; ISBN 0-534-17400-0.

100. Waiblinger, S.; Boivin, X.; Pedersen, V.; Tosi, M.-V.; Janczak, A.M.; Visser, E.K.; Jones, R.B. Assessing the human-animal relationship in farmed species: A critical review. *Appl. Anim. Behav. Sci.* **2006**, *101*, 185–242. [CrossRef]
101. Leach, K.A.; Whay, H.R.; Maggs, C.M.; Barker, Z.E.; Paul, E.S.; Bell, A.K.; Main, D.C.J. Working towards a reduction in cattle lameness: 2. Understanding dairy farmers' motivations. *Res. Vet. Sci.* **2010**, *89*, 318–323. [CrossRef]
102. Farewell dock. Ending tail docking and tail biting in the EU. Available online: http://farewelldock.eu/ (accessed on 4 March 2019).
103. EUWelNet Pig Training. Available online: http://www.euwelnet.eu/en-us/euwelnet-pig-training/ (accessed on 18 March 2019).
104. Teagasc. Environmental Enrichment for Pigs. Available online: https://www.teagasc.ie/media/website/publications/2017/5_Environmental-enrichment-for-pigs.pdf (accessed on 18 March 2019).
105. Harley, S.; Boyle, L.A.; O'Connell, N.E.; More, S.J.; Teixeira, D.L.; Hanlon, A. Docking the value of pigmeat? Prevalence and financial implications of welfare lesions in Irish slaughter pigs. *Anim. Welf.* **2014**, *23*, 275–285. [CrossRef]
106. Sinisalo, A.; Niemi, J.K.; Heinonen, M.; Valros, A. Tail biting and production performance in fattening pigs. *Livest. Sci.* **2012**, *143*, 220–225. [CrossRef]
107. Bolhuis, J.E.; van den Brand, H.; Staals, S.; Gerrits, W.J.J. Effects of pregelatinized vs. native potato starch on intestinal weight and stomach lesions of pigs housed in barren pens or on straw bedding. *Livest. Sci.* **2007**, *109*, 108–110. [CrossRef]
108. Van Dixhoorn, I.D.E.; Reimert, I.; Middelkoop, J.; Bolhuis, J.E.; Wisselink, H.J.; Koerkamp, P.W.G.G.; Kemp, B.; Stockhofe-zurwieden, N. Susceptibility to Co-Infection with Porcine Reproductive and Respiratory Virus (PRRSV) and Actinobacillus pleuropneumoniae (A. pleuropneumoniae) in Young Pigs. *PLoS ONE* **2016**, *11*, e0161832. [CrossRef]
109. Maes, D.; Pluym, L.; Peltoniemi, O. Impact of group housing of pregnant sows on health. *Porc. Heal. Manag.* **2016**, *2*, 17. [CrossRef] [PubMed]
110. Backus, B.L.; McGlone, J.J. Evaluating environmental enrichment as a method to alleviate pain after castration and tail docking in pigs. *Appl. Anim. Behav. Sci.* **2018**, *204*, 37–42. [CrossRef]
111. van Nieuwamerongen, S.E.; Soede, N.M.; van der Peet-Schwering, C.M.C.; Kemp, B.; Bolhuis, J.E. Development of piglets raised in a new multi-litter housing system vs. conventional single-litter housing until 9 weeks of age. *J. Anim. Sci.* **2015**, *93*, 5442–5454. [CrossRef]
112. Aage Arve, N. Heart Pig' Higher Animal Welfare Brand (Denmark). Available online: https://www.eupig.eu/meat-quality/heart-pig (accessed on 18 March 2019).
113. Eskelinen, T.; Räsänen, T.; Santti, U.; Happonen, A.; Kajanus, M. Designing a Business Model for Environmental Monitoring Services Using Fast MCDS Innovation Support Tools. *Technol. Innov. Manag. Rev.* **2017**, *7*, 36–46. [CrossRef]
114. Manning, L. Corporate and consumer social responsibility in the food supply chain. *Br. Food J.* **2013**, *115*, 9–29. [CrossRef]
115. Maloni, M.J.; Brown, M.E. Corporate social responsibility in the supply chain: An application in the food industry. *J. Bus. Ethics* **2006**, *68*, 35–52. [CrossRef]
116. Brinkmann, J. Looking at consumer behavior in a moral perspective? *J. Bus. Ethics* **2004**, *51*, 129–141. [CrossRef]
117. *The Business of Farm Animal Welfare*; Amos, N., Sullivan, R., Eds.; Routledge: Abingdon, UK, 2018; ISBN 978-1-78353-529-3.
118. Sullivan, R.; Elliot, K.; Herron, A.; Vines-Fiestas, H.; Amos, N. Farm animal welfare as an investment issues. In *The Business of Farm Animal Welfare*; Amos, N., Sullivan, R., Eds.; Routledge: Abingdon, UK, 2018; pp. 86–96, ISBN 978-1-78353-529-3.
119. FAIRR. *Factory Farming in Asia: Assessing Investment Risks*. 2017. Available online: https://www.fairr.org/article/report/factory-farming-in-asia-assessing-investment-risks/ (accessed on 9 May 2019).
120. Amos, N.; Sullivan, R. *The Business Benchmark on Farm Animal Welfare Report 2018*. Available online: https://www.bbfaw.com/media/1549/web_bbfaw_report_2018_.pdf (accessed on 6 March 2019).

121. O'Neill, J. *Tackling Drug-Resistant Infections Globally: Final Report and Recommendations*. The Review on Antimicrobial Resistance. 2016. Available online: https://amr-review.org/sites/default/files/160518_Final%20paper_with%20cover.pdf (accessed on 9 May 2019).
122. Lambton, S.L.; Nicol, C.J.; Friel, M.; Main, D.C.J.J.; McKinstry, J.L.; Sherwin, C.M.; Walton, J.; Weeks, C.A. A bespoke management package can reduce levels of injurious pecking in loose-housed laying hen flocks. *Vet. Rec.* **2013**, *172*, 423. [CrossRef] [PubMed]
123. International Finance Corporation. *Improving Animal Welfare in Livestock Operations*. 2014. Available online: https://www.ifc.org/wps/wcm/connect/67013c8046c48b889c6cbd9916182e35/IFC+Good+Practice+Note+Animal+Welfare+2014.pdf?MOD=AJPERES (accessed on 6 March 2019).

© 2019 by the authors. Licensee MDPI, Basel, Switzerland. This article is an open access article distributed under the terms and conditions of the Creative Commons Attribution (CC BY) license (http://creativecommons.org/licenses/by/4.0/).

Commentary

Efforts to Ban the Routine Tail Docking of Pigs and to Give Pigs Enrichment Materials via EU Law: Where Do We Stand a Quarter of a Century on?

Elena Nalon [1],* and Nancy De Briyne [2],*

1. Eurogroup for Animals, Rue Ducale 29, B-1000 Brussels, Belgium
2. Federation of Veterinarians of Europe, Avenue Tervueren 12, 1040 Brussels, Belgium
* Correspondence: e.nalon@eurogroupforanimals.org (E.N.); nancy@fve.org (N.D.B.)

Received: 28 February 2019; Accepted: 25 March 2019; Published: 29 March 2019

Simple Summary: Enforcing legislation on the welfare of pigs is currently one of the European Commission's priorities in the area of animal welfare. This article focuses on the legal ban on the routine docking of tails and the provision of enrichment to pigs within the European Union. It provides a chronological overview of the steps that have been taken by European policy makers to promote the correct implementation and enforcement of the law. In addition, it analyses the current state of play, and presents a reflection on possible future scenarios.

Abstract: In its role as guardian of the Treaties, the European Commission must ensure that Member States enforce EU law within their territories. If adequate enforcement is found to be wanting, the Commission also has the power to take infringement procedures as a corrective measure. The case of Directive 120/2008/EC on the protection of pigs is problematic, as only a few Member States are respecting the ban on routine tail docking, whilst not all pigs are given (adequate) enrichment materials. Twenty-five years after the first EU-wide legal ban on routine tail docking came into force, we are faced with an unprecedented situation that may lead to infringement procedures against more than 20 Member States. This paper describes the various steps that led to the development of the EU law designed specifically to safeguard the welfare of pigs. It lists the numerous efforts (research studies, study visits, recommendations, audits, reports, factsheets, action plans, etc.), undertaken by European decision makers to assist Member States in their efforts to better implement and enforce the relevant rules. Finally, the paper further analyses the current state of play and presents a reflection on possible future scenarios.

Keywords: animal welfare; EU policy; pig directive; enrichment materials; mutilations; straw; swine; tail biting; veterinarian

1. Introduction

Pigs are kept for meat in most European Union (EU) countries. In total in 2017, around 149 million pigs were kept in the EU, with the largest numbers in Spain (31 million), Germany (26 million), France (14 million), Denmark (13 million), the Netherlands (12 million) and Poland (11 million) [1]. Pigs in the EU are mainly kept indoors, with smaller farms having been replaced in the last decade by medium to large-scale farms [2]. Tail biting is a damaging behavior derived from an interplay of factors, some relating to the individual animal concerned (e.g., sex, genetics), others dependent upon how the animals are managed (e.g., lack of manipulable materials, poor climate, feeding problems, dysfunctional social structure, and poor pen layout [3]).

Tail biting is not only an animal welfare problem, as bitten pigs suffer from pain, as well as stress, and can develop infections, but it also carries an economic cost, as such infections can cause carcass condemnations at slaughter [3,4]. However, tail docking per se does not prevent tail biting, and there is evidence that tail biting can be prevented or mitigated in undocked pigs by improving management and housing [3,5].

Under most commercial rearing conditions, the presence of exploration and foraging opportunities is typically minimal or absent. The absence of foraging substrates has been identified as one major, albeit not isolated, risk factor for tail biting [6]. The importance of providing enrichment materials to pigs and reducing stocking densities to manage the risk of tail biting was already stressed in a 1997 report by the Scientific Veterinary Committee, where it stated that *"tail-biting can largely be prevented by providing straw or other manipulable materials and keeping pigs at a stocking density which is not too high."* [7]. In effect, the availability of enrichment materials, especially straw, in animal production systems is widely presumed to be beneficial for the health and welfare of the animals [8]. Pigs in particular have an intrinsic need to explore, even when enough feed is provided. Under semi-natural conditions, domestic pigs have been observed to spend up to 6–8 h foraging for food [9,10]. Foraging activities include rooting, but also grazing and browsing [9]. Exploration of the surroundings is also innate behavior for pigs, occurring even in the absence of external stimuli [10,11]. Appropriate enrichment materials provide pigs with the opportunity to express foraging, rooting, and chewing behaviors, and reduce the occurrence of oral stereotypies [8]. They can also stimulate and satisfy play behavior, which is a very important social aspect to gain social skills, especially in young pigs. Specific edible materials such as straw can supplement fiber to the diet [12] and, if used as bedding, can increase thermal comfort in colder climates. Conversely, it is acknowledged that a lack of environmental enrichment is among the factors determining harmful consequences such as tail biting, ear biting, or aggressive behavior [13,14].

The first EU-wide rules on pig welfare were established in 1991 via Directive 91/630/EEC [15]. This Directive was subsequently amended several times and was substantially updated by Directive 120/2008/EC (henceforth, the Pig Directive [16]). The prohibition on routine tail docking has been in place since 1994, along with the clear stipulation that, to prevent tail biting, enrichment materials such as straw or other suitable materials should be provided to satisfy the behavioral needs of pigs. The Pig Directive specifies the measures that must be undertaken before a farmer can resort to tail docking (i.e., addressing management, stocking density, and providing specific enrichment materials). Nevertheless, tail docking is still practiced routinely in many EU countries [4,17], in violation of these provisions. In Finland and Sweden, due to stricter national rules compared to EU legislation, tail docking is no longer allowed. Outside the EU, in Norway and Switzerland, less than 5% of pigs are tail-docked [4,17]. However, routine tail docking is also carried out in many other countries beyond the Union's borders. Although the current EU pig welfare legislation is in need of updating, as science and technology are constantly evolving, both the enforcement of the provision relating to enrichment and the ban on routine tail docking of pigs have proven to be extremely problematic. In this paper we will examine the EU regulatory background, describe the current situation, and delineate the possible future perspectives on these specific aspects. For an overview at a glance of legislative and non-legislative initiatives leading to the current state of play, see Figure 1.

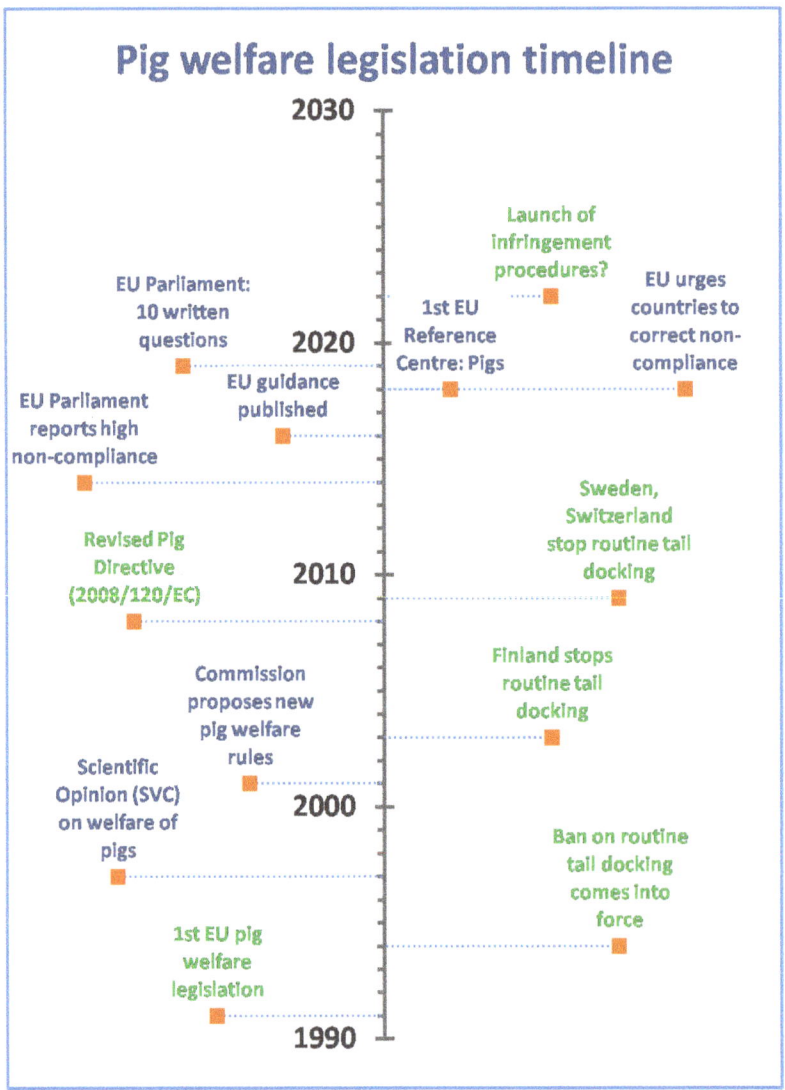

Figure 1. Schematic timeline of EU pig welfare legislation (in green) and non-legislative initiatives (in blue) on the topic of pig welfare with special reference to the ban on routine tail docking.

2. The Process Leading to Current EU Pig Welfare Legislation

2.1. First European Pig Welfare Legislation: 1994

The first EU pig welfare legislation [15] was agreed in 1991. At that time, 12 countries (Belgium, Denmark, France, Germany, Greece, Ireland, Italy, Luxemburg, Netherlands, Portugal, Spain and the UK) were part of the then European Community. The drafting of European animal welfare legislation was initiated as a result of the industrialization of the livestock sector, and the evidence of poor welfare on some farms [18].

It mandated that *"no tail docking must be carried out routinely but only when there is evidence that injuries to other pigs' tails have occurred as a result of not tail docking'* and *'that in addition to the measures normally taken to prevent tail-biting and in order to satisfy their behavioural* (sic) *needs, all pigs, taking into account environment and stocking density, must be able to obtain straw or any other suitable material or objects"*. All member countries had to transpose this Directive into their national legislation, preparing regulations and administrative provisions, including sanctions, by 1 January 1994.

2.2. 1997–2001 Years of Reflection

Article 6 of the 1991 Directive required the Commission to submit a report by 1 October 1997 on intensive pig-rearing systems that complied with the welfare requirements of pigs from the pathological, zootechnical, physiological and behavioral points of view, and on the socio-economic implications of the different systems. The report was delivered by a Scientific Veterinary Committee [7] and provided a detailed analysis of the knowledge (at that time) on the physiology and behavior of domestic pigs and on the health and welfare of intensively kept pigs. More specifically, the report gave 88 recommendations on how the welfare of pigs could be improved, while also taking into account the socio-economic implications. It described in detail the problems seen with tail biting, and a lack of enrichment materials. It confirmed that tail biting should be solved by improving management, rather than tail docking, and stressed the importance of providing enrichment materials to all pigs, and decreasing stocking densities. Four years later, in 2001, the European Commission presented a proposal to the co-legislators, the Council and the European Parliament, stating the intention to revise the 1991 Directive [19,20]. The European Economic and Social Committee (EESC) also presented its opinion on the proposal. Notably the EESC regretted that although the Opinion of the Veterinary Scientific Committee had advised on the minimum space requirements for pigs, the Commission had decided to wait to introduce such rules until the sector was in better financial health. The EESC argued that insufficient space for animals leads to tail biting, meaning that tail docking would remain necessary [21].

2.3. Amended European Pig Directive: 2008

Some aspects of the 1991 Directive were periodically amended, albeit in a piecemeal manner, until the Council of the European Union adopted a substantial revision in 2008. Council Directive 2008/120/EC [16] applied from March 2009, by then covering 27 Member States. Among other important provisions, the Pig Directive reiterated and emphasized the earlier ban on routine tail docking, and the obligation to provide enrichment materials, specifying this time that there should be *"permanent access to a sufficient quantity"* of material to enable proper investigation and manipulation activities. For the first time, the Directive also included a list of materials that can be considered adequate enrichment, namely *"straw, hay, wood, sawdust, mushroom compost, peat or a mixture of such, which does not compromise the health of the animals"* (Annex I, Chapter 1, point 4).

3. 2008–2019: Implementation and (Lack of) Enforcement of Certain Provisions of the Pig Directive

All Member States transposed the Pig Directive into their national law before the legal deadline. Notably in Finland, tail docking had already been banned from 2003, whereas Sweden introduced a total ban through transposing the Directive. These are hitherto the only EU Member States in which demonstrably pig tails are routinely kept intact. In the remainder, it very quickly became clear that the majority (or a large proportion) of pigs continued to be routinely docked, and were not habitually provided with suitable and sufficient enrichment materials.

The widespread failure, not only to enforce the ban on routine tail docking, but also to ensure provision of sufficient and suitable enrichment materials, was brought to the attention of the European Institutions in several ways in the years that followed (for a schematic overview, see Figure 2).

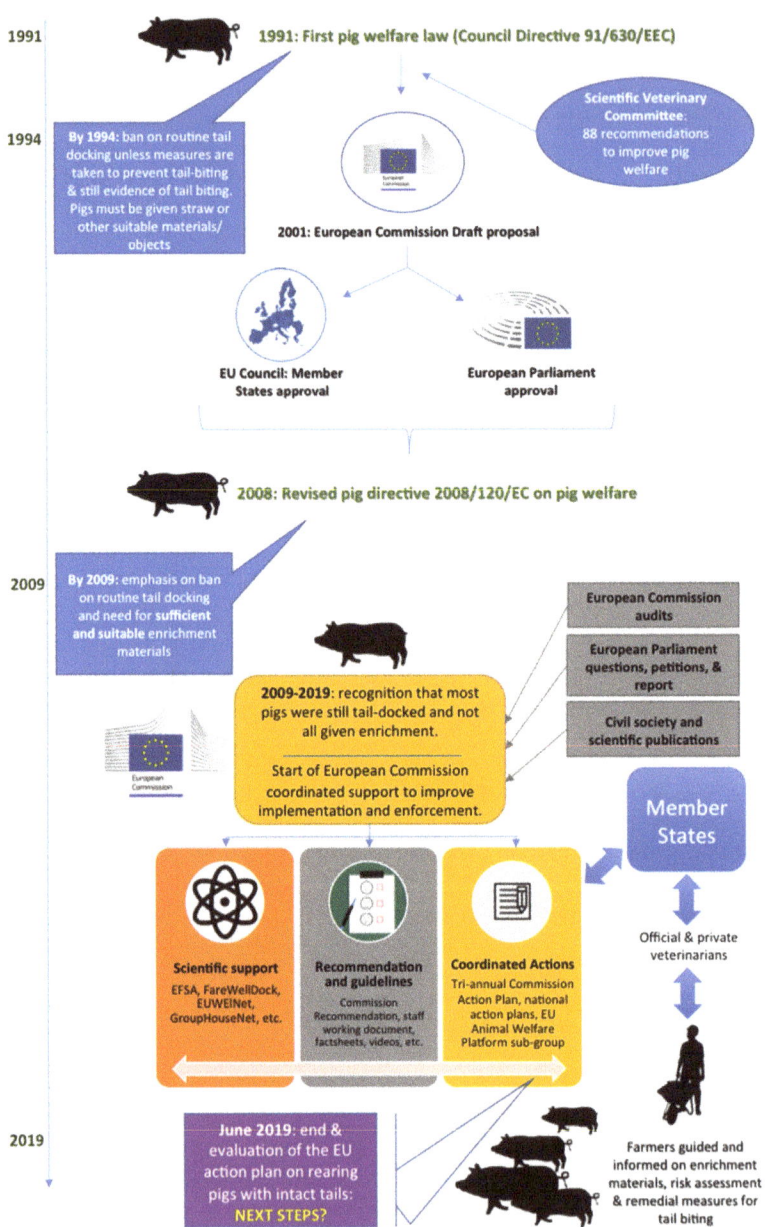

Figure 2. Schematic representation of the temporal sequence of EU-level decision-making processes leading to current pig welfare legislation and actions undertaken by the European Commission to increase Member State compliance with the Pig Directive.

The seventh European Parliament closely scrutinized the lack of implementation of the Pig Directive, with its Members submitting 12 written parliamentary questions to the Commission between 2009 and 2014. A further 10 have been submitted by Members of the eighth European Parliament at the time of writing [22]. In addition, the European Parliament Petitions Committee, a special committee to which all European citizens or residents have the right to send a petition [23] on matters of Union concern, received three petitions on the issue of tail docking and enrichment materials. At the time of writing, these petitions are still open. As a result, in 2014 the Policy Department of the European Parliament prepared a comparative study entitled "In–depth analysis on the routine tail docking of pigs" based on data from Denmark, Sweden, the United Kingdom, Germany, Netherlands and Belgium [24]. This study revealed a *"persisting high rate of non-compliance in the large majority of Member States"* with the EU ban on routine tail docking. It recommended that the European Commission be bolder and adopt *"a stricter enforcement policy"* which, alongside non-legislative tools such as guidelines and e-learning tools for farmers, should also include infringement procedures, considering that *"the mere prospect of serious action may prompt Member States to comply"*. In addition, the European Parliament's Intergroup for the Welfare and Conservation of Animals [25], a cross-group forum open to all Members of the European Parliament with an interest in Animal Welfare, sent multiple letters to the responsible European Commissioners (in both Parliamentary terms), asking them to undertake urgent measures to address the lack of compliance. Subsequent meetings were organized with the respective Commissioners to discuss the way forward, and to request the instigation of infringement procedures.

Serious concerns about the lack of enforcement of the Pig Directive have also been forthcoming from civil society: In 2018, an EU-wide public-facing campaign, coordinated by Eurogroup for Animals [26], collected over one million signatures from European citizens, calling on the European Commission and Member States to fully enforce this law, and to take actions to phase-out the routine surgical castration of piglets without pain relief (for an overview of this issue, see [27]).

Finally, in 2018, the European Court of Auditors issued a report [28] examining the effectiveness of the European Commission's spending on animal welfare measures. The results concerning the Pig Directive are unambiguous: " ... *weaknesses still persisted in some areas related to welfare issues on the farm (in particular, the routine tail docking of pigs"* (p. 5). Additionally, the Court of Auditors found that in two Member States *"Pigs in farms receiving measure 14 support* (i.e., rural development funds for animal welfare under the Common Agricultural Policy) *had their tails docked and did not have access to sufficient enrichment material, as required by the legislation"*. This indicates that some farms that are routinely tail-docking pigs are nonetheless receiving public funds for the purposes of improving animal welfare (measure 14).

The Commission also allocates resources to checking the enforcement of EU legislation. The Directorate General For Health and Food Safety has a specific auditing department (formerly the Food and Veterinary Office, currently Health, Food Audits and Analysis), that verifies that EU legislation on animal health, animal welfare and food safety is properly implemented and enforced [29]. Audits from this unit have confirmed widespread problems with the implementation and enforcement of the ban on routine tail docking, in addition to serious shortcomings concerning the provision of adequate enrichment [30]. To address the situation, an overview report [31] was published in 2016, based on study visits to Finland, Sweden and Switzerland, showcasing good practices for the production of pigs with intact tails, and highlighting why this is not widely practiced in the EU (economics, tail biting and slurry handling in relation to enrichment materials). Nonetheless, the results of the most recent official audits carried out in Germany, the Netherlands, Italy, Spain, and Denmark—the main European pig producing countries—between 2016 and 2018, demonstrate the ubiquitous nature of the problem: 95 to 100% of pigs are still being tail docked [30].

4. Corrective Instruments Available to the European Commission

Member States have the responsibility to implement and enforce EU legislation (directives, regulations, and other legally binding acts). For its part, the European Commission, in its role as the "guardian of the Treaties" and as the Union's executive, has a duty to exert vigilance and ensure proper and timely implementation and enforcement of EU legislation [32]. If one or more Member States fail to transpose and/or enforce EU legislative provisions, the Commission can use various instruments to correct the situation. Such instruments can be broadly classified into two categories: The enforcement approach (coercive) and the management approach (non-coercive), where the Commission has a tradition of associating the two in order to hasten Member State compliance with EU law [32].

The enforcement approach (infringement procedures) is based on the experience that financial sanctions—either threatened or enacted—can deter Member States from ignoring their obligations to enforce Union law, as this becomes damaging both in economic and reputational terms to the Member State concerned [32]. Infringement procedures have three stages: formal notice, reasoned opinion and referral to the Court of Justice (CoJ). The first stage is a *letter of formal notice* by the Commission, requesting information to one or more Member States. If—after receiving the requested information—the Commission finds that the Member State is failing to fulfill its obligations under EU law, it will send a formal request to comply, explaining the reasons why it considers that the country in question is breaching EU law. This second stage is called *reasoned opinion*. The Member State is required to inform the Commission of measures taken to comply within two months. If non-compliance persists, the Commission can initiate the third stage and refer the matter to the Court of Justice (CoJ), and may also ask the CoJ to impose financial penalties (*sanctions*) to the Member State. This stage is called the *Referral to the CoJ*. If the CoJ finds that the Member State has breached EU law, the national competent authorities must take corrective measures.

Financial penalties can also be imposed by the CoJ on the request of the European Commission if, after a first CoJ judgment, a member state still fails to comply with EU legislation. In this case the member state is referred to the CoJ for a second time [33]. Most infringement procedures do not reach the stage of referral to the CoJ, but are resolved earlier, as this system of public "escalation of pressure" is very effective in creating negative publicity for the Member State(s) concerned, and there is a common interest in avoiding costly litigation procedures [33]. Examples of (now closed) infringement procedures started by the Commission in the field of animal welfare include those against Italy for failure to first transpose (Infringement n. 20044171, 2005–2006), and then enforce (Infringement n. 20112231, 2012–2015), the Laying Hen Directive concerning the ban on barren "battery" cages, and that against Greece for failure to enforce the same Directive (Infringement n. 20112230, 2012–2015). Additionally, in February 2013 the Commission sent letters of formal notice to nine Member States (Belgium, Cyprus, Denmark, France, Germany, Greece, Ireland, Poland and Portugal) for failure to address deficiencies in the implementation of the group housing of pregnant sows, which had come into force in January 2013 [34].

The management approach is based on the assumption that Member States must be supported (economically, and with information and guidance) in the adaptations needed to enforce EU legislation [32]. According to some political scientists, the strategy of associating punishment (or the threat of punishment) with persuasive methods has historically proven to be more effective than employing only coercive methods.

5. 2012–2016: A Management-Based Approach to Improve Enforcement of the Pig Directive

In 2016, having noted the unsatisfactory implementation and enforcement of the Pig Directive, the Commission set about remedying this situation, opting for a management-based approach. This approach was carried out in stages, with a period of preparatory actions followed by an operational phase that is still ongoing at the time of writing.

5.1. Scientific Preparatory Work

To increase compliance with the provisions of the Pig Directive and fill knowledge gaps, the Commission funded a series of preparatory studies carried out by scientific consortia (e.g., EUWelNet [35], FareWellDock [36], GroupHouseNet [37]). The studies were aimed at investigating and clarifying practical aspects linked to the management of pigs with intact tails. These projects produced detailed results concerning the effects of different types of enrichment on tail biting, while also taking into account the role of other important factors, such as ventilation, feeding, management, the reduction of stocking densities, and herd health status.

In 2014, the Panel on Animal Health and Welfare (AHAW) of the European Food Safety Authority (EFSA) updated a previous scientific opinion on risk factors for tail biting and possible means to reduce tail docking [17], with new information specifically on manipulable materials and animal-based indicators for pig welfare [6]. The updated scientific review produced guidance on how to practically assess the effectiveness of manipulable materials provided to pigs, as well as the presence and relative importance of risk factors for tail biting. In its conclusions, the EFSA panel found that the presence of manipulable materials is important at all stages of life for the pigs, in order to minimize the risk of undesirable behaviors, including tail biting. The final report stressed the importance of identifying problems and adapting management and environmental factors to control the risk of tail biting. Finally, it recommended two "risk assessment toolboxes" for on-farm use—one to assess the risk of tail biting, and another to assess the adequacy of the enrichment materials provided—that included a combination of resource-based and animal-based indicators.

5.2. Identifying Best Practices and Giving Guidance: The Commission Recommendation and Staff Working Document

On the basis of those preparatory studies, the European Commission set up an expert working group to develop official guidance. The guidance, produced after interservice consultation (consultation within all affected directorate generals of the European Commission), was adopted in 2016. It consisted of two separate documents, the Commission Recommendation 2016/336 [38] "on the application of Council Directive 2008/120/EC laying down minimum standards for the protection of pigs as regards measures to reduce the need for tail-docking", and a Staff Working Document [39] "on best practices with a view to the prevention of routine tail-docking and the provision of enrichment materials to pigs".

The legally binding Commission Recommendation 2016/336 [38] tasks Member States with ensuring that farmers carry out a risk assessment for factors potentially leading to tail biting, and enact corrective measures for each recorded risk factor. It also importantly specifies that enrichment materials must be *"edible, chewable, investigable and manipulable"*, (and explains what these terms mean). Additionally, these materials should be *"of sustainable interest"*, meaning that they should be able to stimulate the exploratory behavior of pigs, and that they should be regularly replaced and replenished. The Recommendation further classifies the conditions that enrichment materials must satisfy to comply with the legal requirements of the Pig Directive. The various admissible materials are ranked based on their adequacy and interest for the animals: optimal materials can be used on their own (examples: straw, fodder), suboptimal materials (examples: wood blocks, ropes) must be used in combination with other materials, whereas materials of marginal interest (examples: plastic balls, metal chains) must be used in combination with optimal or suboptimal materials. The Recommendations also advise farmers and competent authorities to use a combination of resource- and animal-based indicators to assess whether the materials provided are adequate.

The Staff Working Document [39] is meant to give guidance to the pig sector and help competent authorities of Member States understand the rationale behind the Recommendation, while explaining the main scientific principles of pig health and welfare. It provides detailed indications on the types of materials that can be used as enrichment, and on other managerial factors that can help prevent or reduce tail biting (diet, health, thermal comfort, air quality, etc.). The document also provides advice on how to address bouts of tail biting in a herd or group of pigs.

5.3. The 2017–2019 EU Action Plan on Rearing Pigs with Intact Tails

In 2017 the Commission launched a tri-annual EU action plan (2017–2019) based on guidance, the exchange of best practices, study visits, educational materials, stakeholder meetings and consultations, as the main tool to improve compliance with the Pig Directive. The Commission also requested all Member States to provide by December 2018 national action plans detailing how they intend to reach compliance with the requirements of the Directive, with a specific stress on the avoidance of routine tail docking. In 2017, in order to assist Member States, The Directorate-General for Health and Food Safety (DG SANTÉ) produced a series of factsheets [5] covering aspects of pig health and welfare that can contribute to eliminating the need for routine tail docking. The factsheets are accompanied by two videos showing the best practices for rearing pigs with intact tails under completely different systems: the first video shows an example from a big industrial farm in Finland (3000 breeding sows; [40]); the second is about an important Italian producer rearing heavy pigs for the Parma Ham consortium [41]. As pig production systems across the EU vary depending on the geographical region, the library may be expanded to include more examples of best practice from different countries (European Commission, personal communication).

5.4. The European Union Platform on Animal Welfare

In January 2017, following requests from several Member States, the European Commission created the EU Platform on Animal Welfare [42], in order to provide EU and national level agencies with a forum to interact with industry, civil society and academia. The Platform is a Commission expert group that will run until 31 December 2019, and at the time of writing it is not known whether it will be renewed. It brings together 75 institutional, industrial and non-governmental stakeholders, including representatives of all Member States. It has among its main goals to assist the Commission with the development and exchange of coordinated actions to improve enforcement of current animal welfare legislation, which is problematic in several areas, and to this aim it has created several thematic sub-groups. One such sub-group was created in September 2018 to work on pig welfare issues [43]. The objective of the pig sub-group is to advise on how the risk of tail biting can be reduced by meeting the relevant legal requirements in the Pig Directive. Among other activities, the group will also work on animal-based indicators that competent authorities can use to assess the health and welfare status of pigs on farms during official inspections, and which can also be used by quality assurance and agri-food schemes. The first deliverables of the pig sub-group are expected by the end of 2019.

5.5. The first EU Reference Centre on Animal Welfare Dedicated to Pig Welfare

On 5 March 2018 the European Commission designated a first European Union Reference Centre for Animal Welfare [44]. This first Centre, a consortium of three research institutes, is dedicated to pig welfare. Its designation will be reviewed every five years. Its task is to provide technical support and coordinated assistance to the Member States in carrying out official controls in the field of pig welfare and to disseminate good practices. The Centre will also provide scientific and technical expertise, carrying out studies and developing methods for assessing and improving pig welfare. The Centre recently published its first work plan [45] in consultation with EU Member States and other stakeholders and is collaborating with the EU Animal Welfare Platform to communicate results and inform stakeholders on relevant developments.

6. The Future: What Can/Will Happen Next?

2019 will be a year of change within key EU Institutions, not only due to the high probability of the withdrawal of the United Kingdom from the Union, but also as a new European Parliament will be elected, and a new European Commission will be inaugurated. In July 2019, the EU action plan to facilitate the rearing of pigs with intact tails will also end; it will be evaluated and a future approach will need to be decided. The installation of a new executive from the autumn of 2019 will also potentially mean a reshuffling of priorities, and therefore the future approach towards enforcement is currently uncertain. In our opinion, the following scenarios are possible:

1/The new Commission decides to prolong the action plan and continue with a guidance-based approach, still postponing infringement procedures.
2/The new Commission decides to use coercive methods (infringement procedures) with or without a continuation of guidance of Member States to obtain better enforcement.

It must be considered that enforcing EU law is not only a prerogative, but an obligation of the European Commission. It therefore seems implausible that efforts to enforce pig welfare provisions will be disregarded.

7. Discussion

The first ban on routine tail docking across the EU dates from 1994. This means that 1 January 2019 marked the 25th birthday of this provision, which is still largely unenforced by Member States. Whilst infringement procedures may be on the horizon, it is equally plausible that the Commission will only continue to offer guidance and assistance to Member States in their efforts to improve compliance.

Under the current circumstances, the Commission would have to launch over 20 infringement procedures (considering the majority of Member States are non-compliant). However, this is not an exceptional circumstance. By way of an example, as regards air pollution (Directive 2008/50), the Commission declared in 2013 that it had started infringement procedures against 17 Member States for non-compliance with maximum legal limits for PM10 [46]. Until now, the Court of Justice has issued more than 30 judgments against Member States for not complying with the Directive on urban waste water (Directive 91/271), and an equal number with regard to the Directive on nitrates from agricultural production (Directive 91/676). When requested by the Commission, Court of Justice judgments can include financial sanctions that, within the expected spending capacities of the Member State(s), are meant to be conducive to corrective actions (e.g., up to tens of million euros per semester of persistent non-compliance, plus a compensatory bulk sum, and the legal costs). Additionally, the Commission could at least dissuade Member States from some of the worst practices by introducing stricter criteria for obtaining funding under the program for the Promotion of Agricultural Products (part of the budget of the Common Agricultural Policy) managed by the Consumer, Health, Agriculture and Food Executive Agency (CHAFEA). Each year the Commission establishes a line of funding for consortia of European producers to promote "quality products" (products with geographical indication, organic products, etc.) on export markets, and sometimes on the internal market [47]. In recent years, consortia of pig producers located in Member States that—based on the official audit reports by DG SANTÉ—do not respect the Pig Directive regarding the provision of adequate enrichments and the ban on routine tail docking, have been given several million euros in co-funding to promote their pork products abroad. The Commission could, therefore, decide only to give funding to European producers complying completely with the Pig Directive.

What factors have led the European Commission to use a management-based approach over coercive methods? One possible reason is that the EU pig sector is currently faced with the saturation of the demand for pig meat on the internal market, coupled with an animal health crisis due to African Swine Fever (ASF). ASF is a highly contagious viral disease that affects wild boars and domestic pigs, with a high degree of morbidity and mortality, and which disrupts exports and leads to substantial economic losses [48]. It should be noted that the EU has a self-sufficiency of about 111% for pig meat, and exports about 13% of its total production. The EU is the first global exporter of pig meat worldwide [49] and arguably wants to maintain this position.

The European livestock sector is concerned that EU animal welfare legislation is causing a competitive disadvantage for EU products on the global market [50,51]. In 2018, to respond to these concerns, the Commission published a 'Report on the impact of animal welfare international activities on the competitiveness of European livestock producers in a globalized world' [52]. The report concluded that the EU has played a prominent and decisive role in raising global awareness on animal welfare, and that significant results have been achieved. The report also found that animal welfare standards have a limited impact on the competitiveness of EU producers on world markets: The overall costs of compliance with animal welfare standards remains very low compared to other production costs (such as the cost of labor and feed) that affect global competitiveness and influence world trade patterns. Therefore, EU animal welfare standards can even offer an opportunity to better valorize the added market value of EU products. However, specifically on the issue of stopping routine tail docking and providing enrichment materials, farmers are reluctant to implement changes. This can be explained through three main hypotheses: (i) due to the multifactorial nature of the problem, it may be difficult to predict and completely prevent episodes of tail biting; (ii) habit, i.e., the tendency to repeat habitual behavior, and limited knowledge about the alternatives; (iii) raising pigs with intact tails costs more, as a result of the need to decrease stocking density, provide more enrichment material, and the labor costs involved, including increased vigilance on the animals to identify (and remove, if necessary) those that have a tendency to bite [4]. The official audit reports of the Commission from Italy [53] and Spain [54] confirm this, and show that farmers consider it very difficult (if not impossible) to rear pigs with intact tails in the existing systems without giving pigs more space and more enrichment. They also consider that such improvements are not realistic under their countries' commercial rearing conditions.

On the other hand, farmers from countries with very different rearing systems are increasingly finding solutions that work for them. Successful examples of rearing pigs with intact tails have been reported from countries such as Germany, Denmark, Italy, and Ireland [55–57], as well as Finland and Sweden. Advantages reported by farmers include an increased pride in their work, an improvement in animal health with reduced need for antibiotics, and the possibility to obtain better prices on the market. The common denominator of these experiences is a substantial change of management, in which careful stockmanship, lower stocking densities, improved air and water quality, and the provision of a stress-free environment, play a major role. These factors are clearly summarized in the Commission's online resources [5].

Member States are responsible for ensuring timely transposition and enforcement of EU law on their territories. There are various disincentives that can lead to lack of compliance, including financial costs, cost-benefit considerations, pressure from lobbies, and a lack of administrative infrastructure [32,58]. The official audits of the Commission stressed the lack of information in respect to the proportion of pigs that are tail docked, and the percentage of tail biting in different Member States. Measuring is a pre-condition for improvement. It is crucial to set up an effective monitoring program, both on the farm and at the abattoir, to record tail biting lesions and the percentage of docked pigs. These aspects will be addressed by the EU Reference Centre on Pig Welfare and within the dedicated sub-group of the EU Animal Welfare Platform. Developing reliable and harmonized animal-based indicators to measure animal welfare and assess compliance with legislative requirements has become a priority. Apart from drafting risk assessment templates, action plans to address identified risk factors, and taking controls seriously, Member States also have at their disposal a powerful instrument to help farmers financially to improve pig welfare, and eventually stop routine tail docking: Common Agricultural Policy spending under Pillar II. The rural development Regulation [59] provides Member States with possibilities to support investments specifically dedicated to animal welfare: so far these funds have been primarily used for the "modernization" of buildings, but incentives could also be given for reducing stocking densities, improving air and water quality, providing outdoor access where feasible, keeping animals on straw bedding where climate permits, or changing manure disposal systems so that straw can be provided as an enrichment.

Last, but not least, the role of civil society in determining the speed of change cannot be underestimated. It is also thanks to the petitions initiated in the European Parliament by private citizens, the tenacious actions of Members of the European Parliament, the investigative materials, reports and signature collections received from non-governmental organizations, that the Commission has taken more decisive action. Indeed, external pressure (via the so-called "societal watchdogs") is one of the fundamental instruments that the Commission uses to compensate for the lack of internal resources, when checking if EU legislation is being implemented and enforced [32].

To conclude, all the relevant actors—the Commission, the Member States, private and official veterinarians, farmers, and civil society at large—will have a role to play in increasing the chance of enforcing these 25-year old pig welfare provisions. The implications of keeping pigs with intact tails and providing them with meaningful and adequate enrichment are of course wider than a discussion about the purely technical solutions. Farm animal welfare is of paramount importance for European citizens, as shown by the results of the last special Eurobarometer on Animal Welfare [60], and scientific evidence is increasingly confirming the highly developed cognitive and emotional capabilities of farm animals, and the relevance of providing them with "lives worth living" [61]. Any long-standing and widespread lack of enforcement of EU legislation is thus detrimental not only for the animals, but also for the reputation of EU law as driver of better farm animal welfare globally [18]. A looming question remains unaddressed, and very difficult to answer. What if all efforts should fail? The consequences could potentially be profound, and call for a serious reflection on our current animal farming paradigm.

Author Contributions: Conceptualization, E.N. and N.D.B.; resources, E.N. and N.D.B.; writing—original draft preparation, E.N. and N.D.B.; writing—review and editing E.N. and N.D.B.

Funding: This research received no external funding.

Acknowledgments: The authors would like to thank Reineke Hameleers and Jan Vaarten for reading and giving comments on the draft text. They would also like to wholeheartedly thank Joe Moran for English language revision.

Conflicts of Interest: The authors declare no conflict of interest.

References

1. European Commission. Eurostat. Pig Population/Livestock Survey—Annual Data. 2019. Available online: https://ec.europa.eu/agriculture/sites/agriculture/files/market-observatory/meat/pigmeat/doc/pig-population-survey_en.pdf (accessed on 23 February 2019).
2. European Commission. Eurostat. Statistics Explained: Pig Farming Sector Statistical Report. 2014. Available online: https://ec.europa.eu/eurostat/statistics-explained/index.php/Archive:Pig_farming_sector_-_statistical_portrait_2014 (accessed on 23 February 2019).
3. Valros, A.; Heinonen, M. Save the pig tail. *Porc. Health Manag.* **2015**, *1*, 2. [CrossRef] [PubMed]
4. De Briyne, N.; Berg, C.; Blaha, T.; Palzer, A.; Temple, D. Phasing out Pig Tail Docking in the EU—Present State, Challenges and Possibilities. *Porc. Health Manag.* **2018**, *4*, 1–9. [CrossRef] [PubMed]
5. European Commission. Cutting the Need for Tail-Docking. 2016, p. 14. Available online: https://ec.europa.eu/food/sites/food/files/animals/docs/aw_practice_farm_pigs_tail_docking_eng.pdf (accessed on 28 February 2019).
6. European Food Safety Authority EFSA, AHAW Panel. Scientific Opinion concerning a Multifactorial approach on the use of animal and non-animal-based measures to assess the welfare of pigs. *EFSA J.* **2014**, *12*, 3702.
7. European Union. The Welfare of Intensively Kept Pigs. Report of the Scientific Veterinary Committee 1997. Available online: https://ec.europa.eu/food/sites/food/files/animals/docs/aw_arch_1997_intensively_kept_pigs_en.pdf (accessed on 18 March 2019).
8. Tuyttens, F.A.M. The Importance of Straw for Pig and Cattle Welfare: A Review. *Appl. Anim. Behav. Sci.* **2005**, *92*, 261–282. [CrossRef]
9. Wood-Gush, D.G.M.; Jensen, P.; Algers, B. Behaviour of pigs in a novel seminatural environment. *Biol. Behav.* **1990**, *15*, 62–73.
10. Studnitz, M.; Jensen, M.B.; Pedersen, L.J. Why do pigs root and in what will they root? A review on the exploratory behaviour of pigs in relation to environmental enrichment. *Appl. Anim. Behav. Sci.* **2007**, *107*, 183–197. [CrossRef]
11. Wood-Gush, D.G.M.; Vetergaard, K. Inquisitive exploration in pigs. *Anim. Behav.* **1993**, *45*, 185–187. [CrossRef]
12. Bindelle, J.; Leterme, P.; Buldgen, A. Nutritional and Environmental Consequences of Dietary Fibre in Pig Nutrition: A Review. *Biotechnol. Agron. Soc. Environ.* **2008**, *12*, 69–80.

13. Sonoda, L.; Fels, M.; Oczak, M.; Vranken, E.; Ismayilova, G.; Guarino, M.; Viazzi, S.; Bahr, C.; Berckmans, D.; Hartung, J. Tail Biting in pigs—Causes and management intervention strategies to reduce the behavioural disorder. A review. *Berl. Munch. Tierarztl. Wochenschr.* **2013**, *126*, 104–112. [PubMed]
14. Valros, A.; Munsterhjelm, C.; Hänninen, L.; Kauppinen, T.; Heinonen, M. Managing undocked pigs – on-farm prevention of tail biting and attitudes towards tail biting and docking. *Porc. Health Manag.* **2016**, *2*, 2. [CrossRef]
15. European Union. Council Directive 91/630/EEC of 19 November 1991 laying down minimum standards for the protection of pigs. *Off. J. Eur. Union* **1991**, *50*, 33–38.
16. European Union. Council Directive 2008/120/EC of 18 December 2008 laying down minimum standards for the protection of pigs (Codified version). *Off. J. Eur. Union* **2009**, *L47*, 5–13.
17. European Food Safety Authority EFSA. The risks associated with tail biting in pigs and possible means to reduce the need for tail docking considering the different housing and husbandry systems. *EFSA J.* **2007**, *5*, 611. [CrossRef]
18. European Parliament. Directorate General for Internal Policies. Policy Department C—Citizens' Rights and Constitutional Affairs. Animal Welfare in the European Union—Study for the Peti Committee. PE 583.114. 2017. Available online: http://www.europarl.europa.eu/RegData/etudes/STUD/2017/583114/IPOL_STU(2017)583114_EN.pdf (accessed on 18 March 2019).
19. European Commission. Communication from the Commission to the Council and the European Parliament on the Welfare of Intensively Kept Pigs in Particularly Taking into Account the Welfare of Sows Reared in Varying Degrees of Confinement and in Groups. COM/2001/0020. 2001. Available online: https://eur-lex.europa.eu/legal-content/EN/TXT/HTML/?uri=COM:2001:20:FIN&from=EN (accessed on 23 February 2019).
20. European Commission. Proposal for a Council Directive amending Directive 91/630/EEC laying down minimum standards for the protection of pigs. *Off. J. Eur. Union* **2001**, *C 154 E/114*, P0114–P0116.
21. European Commission. Opinion of the Economic and Social Committee. *Off. J. Eur. Commun.* **2001**, *C221*, 74.
22. European Parliament Plenary, Written Questions, Parliamentary Terms 2009–2014 and 2014–2019. Available online: http://www.europarl.europa.eu/plenary/en/parliamentary-questions.html?tabType=wq#sidesForm (accessed on 28 February 2019).
23. European Parliament. Petitions. Available online: http://www.europarl.europa.eu/at-your-service/en/be-heard/petitions (accessed on 18 March 2019).
24. Marzocchi, O. Routine Tail Docking of Pigs. Study for the PETI Committee. 2014. European Parliament, Directorate-General for Internal Policies. Policy Department C: Citizens' Rights and Constitutional Affairs. Available online: http://www.europarl.europa.eu/RegData/etudes/STUD/2014/509997/IPOL_STU%282014%29509997_EN.pdf (accessed on 18 March 2019).
25. The European Parliament's Intergroup on the Welfare and Conservation of Animals. Working to Ensure That Full Regard Is Paid to the Welfare Requirements of Animals. Available online: http://www.animalwelfareintergroup.eu/ (accessed on 18 March 2019).
26. End Pig Pain. Available online: www.endpigpain.eu (accessed on 28 February 2019).
27. De Briyne, N.; Berg, C.; Blaha, T.; Temple, D. Pig Castration: Will the EU Manage to Ban Pig Castration by 2018? *Porc. Health Manag.* **2016**, *2*, 1–11. [CrossRef]
28. European Court of Auditors. Animal Welfare in the EU: Closing the Gap between Ambitious Goals and Practical Implementation. 2018, 31, p. 68. Available online: https://www.eca.europa.eu/Lists/ECADocuments/SR18_31/SR_ANIMAL_WELFARE_EN.pdf (accessed on 18 March 2019).
29. European Commission, Directorate Health and Food Audits and Analysis. General Page. Available online: https://ec.europa.eu/food/audits_analysis_en (accessed on 28 February 2019).
30. European Commission, Directorate Health and Food Audits and Analysis. Audit Reports. Available online: http://ec.europa.eu/food/audits-analysis/audit_reports/index.cfm (accessed on 28 February 2019).
31. European Commission, Directorate Health and Food Audits and Analysis. Overview Report Study Visits on Rearing Pigs with Intact Tails. 2016. Available online: http://ec.europa.eu/food/audits-analysis/overview_reports/act_getPDF.cfm?PDF_ID=790 (accessed on 18 March 2019).
32. Tallberg, J. Paths to Compliance: Enforcement, Management, and the European Union. *Int. Organ.* **2002**, *56*, 609–643. [CrossRef]

33. European Commission, Policies, Information and Services. Infringement Procedure. Available online: https://ec.europa.eu/info/law/law-making-process/applying-eu-law/infringement-procedure_en (accessed on 28 February 2019).
34. European Commission Press Release. Animal Welfare: Commission Increases Pressure on Member States to Enforce Group Housing of Sows. 2013. Available online: http://europa.eu/rapid/press-release_IP-13-135_en.pdf (accessed on 28 February 2019).
35. EUWelNet. Understanding Environmental Enrichment and Tail Docking Requirements for Finisher Pigs in Accordance with EU Directive 2008/120/EC. 2013. Available online: http://pigstraining.welfarequalitynetwork.net/Pages/0 (accessed on 28 February 2019).
36. Bracke, M.B.M.; Koene, P. Did European Pig-Welfare Legislation Reduce Pig Welfare? Perhaps Not, but Experts Confirm that Common Indestructible Materials Are Not Proper Enrichment for Pigs at All, Except Perhaps for an Enhanced Novel Branched-Chains Design. FareWellDock, 2019. Available online: https://farewelldock.eu/ (accessed on 23 February 2019).
37. GroupHouseNet. Synergies for Preventing Damaging Behaviour in Group Housed Pigs and Chickens. COST Action CA15134. Available online: https://www.grouphousenet.eu/ (accessed on 15 March 2019).
38. European Union. Commission Recommendation (EU) 2016/336 on the application of Council Directive 2008/120/EC laying down minimum standards for the protection of pigs as regards measures to reduce the need for tail-docking. *Off. J. Eur. Union* **2016**, *L62*, 20.
39. European Commission. Commission Staff Working Document on Best Practices with a View to the Prevention of Routine Tail-Docking and the Provision of Enrichment Materials to Pigs. 2016. Available online: https://ec.europa.eu/food/sites/food/files/animals/docs/aw-pract-farm-pigs-staff-working-document_en.pdf (accessed on 15 March 2019).
40. European Commission, Audiovisual Services. Animal Welfare/Pigs/Tail-Docking, Finland. Available online: http://ec.europa.eu/avservices/video/player.cfm?ref=I147131&sitelang=en&lg=FI/EN (accessed on 28 February 2019).
41. European Commission, Audiovisual Services. Animal Welfare/Pigs/Tail-Docking, Italy. Available online: http://ec.europa.eu/avservices/video/player.cfm?ref=I147129&sitelang=en&lg=IT/EN (accessed on 28 February 2019).
42. European Commission. Commission Decision of 24 January 2017 Establishing the Commission Expert Group 'Platform on Animal Welfare'. *Off. J. Eur. Union* **2017**, *C31*, 61.
43. European Commission. EU Platform on Animal Welfare. Thematic Sub-Groups: Pigs. Available online: https://ec.europa.eu/food/animals/welfare/eu-platform-animal-welfare/thematic-sub-groups/pigs_en (accessed on 28 February 2019).
44. Commission Implementing Regulation (EU) 2018/329 of 5 March 2018 Designating a European Union Reference Centre for Animal Welfare. *Off. J. Eur. Union* **2018**, *L63*, 13.
45. Spoolder, H. EU Reference Centres for Animal Welfare. Work Programme. 2018. Available online: https://ec.europa.eu/food/sites/food/files/animals/docs/aw_eu-reference-centre_work-prog.pdf (accessed on 28 February 2019).
46. European Commission. Press Release. Environment: A Fresh Legal Approach to Improving Air Quality in Member Stats. 2013. Available online: http://europa.eu/rapid/press-release_IP-13-47_en.htm (accessed on 15 March 2019).
47. CHAFEA, Promotion of Agricultural Products. Campaigns Maps and Statistics. Available online: https://ec.europa.eu/chafea/agri/campaigns/map-and-statistics-target-countries (accessed on 28 February 2019).
48. European Food Safety Authority. African Swine Fever. 2018. Available online: https://www.efsa.europa.eu/en/topics/topic/african-swine-fever (accessed on 15 March 2019).
49. European Commission, Agriculture and Rural Development, EU Market Observatories, EU Meat Market Observatory. Pig Meat. Available online: https://ec.europa.eu/agriculture/sites/agriculture/files/market-observatory/meat/pigmeat/doc/pig-market-situation_en.pdf (accessed on 28 February 2019).
50. Copa-Cogeca. The Reaction of European Farmers and European Agri-Cooperatives to the EU Strategy for the Protection and Welfare of Animals 2012–2015. 2015. Available online: http://copa-cogeca.eu/Download.ashx?ID=938718 (accessed on 18 March 2019).

51. European Commission, Directorate-General for Agriculture and Rural Development. Final Report: Assessing Farmers' Cost of Compliance with EU Legislation in the Fields of Environment, Animal Welfare, and Food Safety. 2014. Available online: https://ec.europa.eu/agriculture/sites/agriculture/files/external-studies/2014/farmer-costs/fulltext_en.pdf (accessed on 28 February 2019).
52. European Commission. Report from the EC to the EP and the Council on the Impact of Animal Welfare International Activities on the Competitiveness of European Livestock Producers in a Globalized World. 2018. Available online: https://ec.europa.eu/food/sites/food/files/animals/docs/aw_international_publication-report_en.pdf (accessed on 18 March 2019).
53. European Commission, Directorate General for Health and Food Safety—Health and Food Audits and Analysis. Final Report of an Audit Carried out in Italy from 13 November 2017 to 17 November 2017 in Order to Evaluate Member State Activities to Prevent Tail-Biting and Avoid Routine Tail-Docking of Pigs. DG SANTE 2017-6257. Available online: http://ec.europa.eu/food/audits-analysis/act_getPDF.cfm?PDF_ID=13722 (accessed on 28 February 2019).
54. European Commission, Directorate General for Health and Food Safety—Health and Food Audits and Analysis. Final Report of an Audit Carried out in Spain from 18 September 2017 to 22 September 2017 in Order to Evaluate Member State Activities to Prevent Tail-Biting and Avoid Routine Tail-Docking of Pigs. DG SANTE 2017-6126. Available online: http://ec.europa.eu/food/audits-analysis/act_getPDF.cfm?PDF_ID=13649 (accessed on 28 February 2019).
55. van Bebber, J. Rearng Pigs with Intact Tails (Hof Bodenkamp, Germany): Assessing and Addressing Risks. Meeting of the Action Plan of the European Commission on Rearing Pigs with Intact Tails. 2017. Available online: https://circabc.europa.eu/sd/a/888d4c2b-cf2c-48aa-a6e0-81e1d08d434e/Rearing%20pigs%20with%20intact%20tails%20in%20EU%20(DE)%20-%20assessing%20and%20addressing%20risks_VAN%20BEBBER%20J_2017_EN(0).pdf (accessed on 28 February 2019).
56. Aage Arve, N. Rearing Pigs with Intact Tails (Krannestrup, Denmark). Meeting of the Action Plan of the European Commission on Rearing Pigs with Intact Tails. 2017. Available online: https://circabc.europa.eu/sd/a/f5524714-2e16-46cd-98fd-b01c66498b43/Rearing%20pigs%20with%20intact%20tails%20in%20EU%20(DK)%20-%20assessing%20and%20addressing%20risks_AAGE%20ARVE%20N_2017_EN%20.ppt (accessed on 28 February 2019).
57. Pisapia, A. Italian Ministry's Action Plan to Avoid Tail Docking. CIWF's Perspective. Meeting of the Action Plan of the European Commission on Rearing Pigs with Intact Tails. 2018. Available online: https://circabc.europa.eu/sd/a/eb17acd6-2fc2-4126-9a45-ea78594d5614/Working%20together%20towards%20rearing%20pigs%20with%20intact%20tails%20in%20Italy%20(CIWF)_Pisapia%20A_2018_EN.ppt (accessed on 28 February 2019).
58. Börzel, T.A.; Buzogány, A. Compliance with EU environmental law. The iceberg is melting. *Environ. Polit.* **2019**, *28*, 315–341. [CrossRef]
59. Regulation (EU) No 1305/2013 of the European Parliament and of the Council of 17 December 2013 on Support for Rural Development by the European Agricultural Fund for Rural Development (EAFRD) and Repealing Council Regulation (EC) No 1698/2005. *Off. J. Eur. Union* **2013**, *L347*, 487–548.
60. European Commission. Special Eurobarometer 442. Report: Attitudes of Europeans towards Animal Welfare. 2016. Available online: http://ec.europa.eu/COMMFrontOffice/publicopinion/index.cfm/ResultDoc/download/DocumentKy/71348 (accessed on 28 February 2019).
61. Mellor, D. Updating animal welfare thinking: Moving beyond the "Five Freedoms" towards "a Life Worth Living". *Animals* **2016**, *6*, 21. [CrossRef] [PubMed]

© 2019 by the authors. Licensee MDPI, Basel, Switzerland. This article is an open access article distributed under the terms and conditions of the Creative Commons Attribution (CC BY) license (http://creativecommons.org/licenses/by/4.0/).

Article

Tail Posture as an Indicator of Tail Biting in Undocked Finishing Pigs

Torun Wallgren *, Anne Larsen and Stefan Gunnarsson

Department of Animal Environment and Health, Swedish University of Agricultural Sciences (SLU), PO-Box 234, 532 23 Skara, Sweden; Anne.Larsen@slu.se (A.L.); Stefan.Gunnarsson@slu.se (S.G.)
* Correspondence: Torun.Wallgren@slu.se; Tel.: +46-511-672-19

Received: 5 November 2018; Accepted: 3 January 2019; Published: 8 January 2019

Simple Summary: Tail biting is a large welfare problem in modern pig production, causing pain and reduced health and production. The identification of tail biting is important for minimising the risk of the escalation of the behaviour and its consequences. Tail posture (i.e., tail hanging or curled) has been suggested to depend on the presence of tail wounds and, therefore, has been suggested as an indicator of tail biting. This study investigated the relationship between tail position and tail damages at feeding, since that could be a feasible time for producers to detect tail posture. The experiment showed that 94% of the pigs had curly tails and that pigs with wounds were more likely to have hanging tails than pigs with nondamaged tails. By observing the tail position at feeding, we were able to identify pigs with tail wounds in 68% of cases simply by scoring pigs with hanging tails. To conclude, the scoring of pigs with hanging tails at feeding was found to be a useful tool for identifying tail damages, which may otherwise be difficult to detect by the caretaker.

Abstract: Tail posture (i.e., hanging or curled) has been suggested to be an indicator of tail biting, and hanging tails predisposed to damage. The aim of this study was to investigate if tail posture was feasible as a tail damage indicator in a commercial setting. The study was carried out on one batch of 459 undocked finishing pigs (30–120 kg in weight). Weekly scoring of tail posture was combined with the scoring of tail lesions. Tail posture was observed at feeding to facilitate the usage of the method in commercial settings. A curly tail was observed in 94% of the observations. Pigs with tails scored with "wound" were 4.15 ($p < 0.0001$) times more likely to have hanging tails, and pigs scored with "inflamed wounds" were 14.24 ($p < 0.0001$) times more likely to have hanging tails, compared to pigs with nondamaged tails. Tail posture correctly classified tails with "wound" or "inflamed wound" 67.5% of the time, with 55.2% sensitivity and 79.7% specificity, respectively. The method of observing the tail position at feeding seems useful as a complement to normal inspection for detecting tail biting before tail wounds are visible to the caretaker.

Keywords: tail docking; animal welfare; swine; fattening pig; tail damage

1. Introduction

Tail biting is a well-known issue within European pig production. In this context, tail biting is defined as one pig orally manipulating another pig's tail, and the phenomenon occurs in both docked and undocked pig populations [1]. Tail biting and subsequent tail injuries are known to cause stress and reduce welfare in both the injured and biting pigs [2]. Additionally, tail wounds may reduce weight gain and cause condemnation of the whole or parts of the pig carcass. Therefore, the prevention of tail biting is important for profitability, as well as the improvement of animal welfare. Although both legislative and consumer demands require pigs to be raised without tail docking [3], the most recent survey published by the EFSA council (2007) shows that 90–95% of the pigs produced within the EU are tail docked to reduce the risk and consequences of tail biting [1,4,5]. Raising undocked pigs has been associated with an increased risk of tail biting, and therefore management routines for raising pigs with intact tails must be developed before realistically eliminating tail docking within the EU [4,6].

Even though a lack of occupation has been found to be one of the major risk factors for developing tail biting in pigs [1], the causal background of tail biting outbreaks is complex and has not yet been fully understood. It has been suggested that tail biting is a redirected behaviour with its background in unfulfilled exploratory behaviour [7]. Nevertheless, tail biting is multifactorial, with several other factors having an impact on the development of tail biting, such as genetics, feed type and indoor climate [2,8]. It is hard to stop an outbreak once tail biting has developed [9]. Fraser [10] found that pigs show an increased attraction to chewing the tails of other pigs when blood is present, although this attraction to blood is highly individual. This phenomenon is considered part of the explanation why small tail wounds may lead to large tail biting outbreaks within a short period of time [10].

Since tail biting outbreaks may escalate rapidly and are difficult to stop, the emphasis must focus on minimizing the risk factors of outbreaks. It has been hypothesised that if tail biting behaviour is detected early, i.e., before severe tail wounds appear, a change in management might inhibit an outbreak. Tail biting behaviour is usually not detected until tail lesions are present, which increases the difficulties in stopping outbreaks [11]. Tail posture has previously been suggested to be an indicator of tail biting in pigs, with researchers hypothesising that affected pigs should be more prone to have their tail in a hanging posture than unaffected pigs [12,13].

Already in 1990, McGlone et al. [14] noted pigs having uncurled tails when tail biting occurred. Tail posture has been suggested to be a protective measure as well as being part of pig communication and comfort [12,14]. For example, heat-stressed pigs may be more likely to have hanging tails, whereas pigs being handled by a familiar person, experiencing positive pig comfort, might have curled tails [12]. In weaned piglets, it has been shown that tail posture is related to tail lesions and that hanging tails may be a predictor for tail lesions occurring 2–3 days later [11]. According to Lahrmann et al. [15], hanging tails were more common in pigs kept in pens where tail biting had been observed than on pigs in pens where no tail biting had been observed. They also found that the number of pigs with hanging tails observed on one day was correlated to the number of tail lesions observed the day after. A recent study by Larsen et al. [13] found that a tucked-in tail increased the odds for tail wounds on the same day in finishing pigs. However, their method was found to give many false identifications. Therefore, we hypothesised that tail posture could be a suitable option for the early detection of tail biting under commercial conditions due to the ease of recognition of curled or hanging tails.

The aim of this study was to investigate the relationship between tail posture (hanging or curled) at feeding and tail lesions in finishing pigs. The hypothesis was that pigs with hanging tails were predisposed to having their tails bitten and had tail lesions that could be assessed with closer visual examination. It was further hypothesised that tail posture would be easily detectable at feeding, which could then be used as an intervention method at commercial farms. Furthermore, we attempted to determine the specificity and sensitivity of tail posture observation in relation to the scoring of tail lesions.

2. Materials and Methods

The study was carried out at a commercial farrow-to-finish pig farm in southwestern Sweden for a total of 102 days. A total of 14 observations were carried out (one per week) from December 2017 to March 2018. One batch of 458 pigs was studied from approximately 30 kg live weight (LW) until slaughter (approximately 120 kg LW). Pigs were sent to slaughter in five shipments based on LW, and the trial was ended when the majority (i.e., >70%) of the pigs had been sent to slaughter, on week 14 (Table 1). All pigs were the progeny of crossbred sows, Landrace*Yorkshire (TN70) and Hampshire boars. All sows were kept in loose-housed farrowing systems with straw. All pigs were undocked and males were surgically castrated by two incisions made in the scrotum during the first week of life after receiving analgesic treatment. The analgesia was performed with an injection of 0.3–0.5 mL/testicel of lidocaine 20 mg/mL and adrenalin 0.036 mg/mL (Lidokel-Adrenalin vet.®). The piglets were weaned at approximately five weeks of age.

Table 1. Descriptive data of tail damage score in relation to tail posture, presented over the production weeks. Tail posture: C = curled tail and H = hanging tail. Pigs for which no observation was made for tail posture or tail damage were omitted from further observations; consequently, 'Number of obs.' and 'No of pigs' may differ. Pigs were sent to slaughter in weeks 10, 12 and 14.

Week	Tail Posture	Damage				Total	Number of Obs.	Number of Pigs
		Nondamaged	Swollen	Bite Mark	Wound	Inflamed Wound			
1	C	315	41	32	40	1	429	457	459
	H	14	2	2	8	2	28		
2	C	315	41	38	52	3	449	457	459
	H	2			5	1	8		
3	C	281	41	42	50	1	415	451	457
	H	17	2	1	14	2	36		
4	C	252	33	66	58	0	409	449	453
	H	22	4		12	2	40		
5	C	250	46	51	85	3	435	447	452
	H	3	2	1	5	1	12		
6	C	214	47	83	77	4	425	449	452
	H	7	1	4	10	2	24		
7	C	202	42	81	95	3	423	446	450
	H	9	1	2	11		23		
8	C	199	54	64	112	3	432	444	449
	H		2	1	7	2	12		
9	C	238	27	80	79	5	429	445	448
	H	4	1		9	2	16		
10	C	206	66	44	84	3	403	445	447
	H	8	6	5	20	3	42		
11	C	179	47	60	120	5	411	430	435
	H	3		2	12	2	19		
12	C	186	49	46	96	3	380	428	433
	H	7	4	6	28	3	48		
13	C	18	35	40	79	5	337	370	378
	H	5	2	3	20	3	33		
14	C	148	40	49	90	9	336	374	378
	H	6	5	2	17	8	38		

When moved into the finishing pig stable, pigs were sorted by size (largest pigs kept together and smallest pigs kept together) but not by sex. The 458 pigs studied were allocated into 42 pens, which housed either 10 (n = 4), 11 (n = 37) or 12 (n = 1) pigs per pen. The pens each had an area of 10.49 m^2, 7.81 m^2 of which was solid floor and 2.68 m^2 slatted floor. In each pen there was a 3.4 m-long feeding trough at the long side of the pen and a nipple drinker above the slatted area. The pigs were fed liquid feed with an automatic feeding system according to the farm feeding regime of four meals per day until week 12, when the feeding regime was changed to three meals per day. The pigs were inspected daily and the pens were manually cleaned and provided with fresh chopped straw once a day (~25 L of straw provided on the floor or ~44 L provided in a straw rack; 25 L of straw weighs ~1.8 kg).

The pigs were not individually tagged; therefore, in order to keep track of individuals, they were marked with spray paint (PORCIMARK marking spray, Kruuse, Denmark) twice per week. The animals were marked with one to three lines on their back in red, blue or green. One pig per pen was kept unmarked.

The tails of the pigs were scored weekly by palpation with regard to tail shortening, tail damage, and wound freshness, according to the scoring protocol presented in Table 2. Pigs that were observed to be limping severely or unwilling to stand or put weight on at least one of their limbs at scoring were recorded as lame. All scoring was carried out by the same operator.

Table 2. Tail lesion scoring of tail length, tail damage and wound freshness, adapted from Zonderland et al. [16].

Tail Lesion	Score 0	Score 1	Score 2	Score 3	Score 4
Length	Not shortened	Shortened	Short		
	No shortening	A part of the tail tissue has been bitten off and the tail has been shortened to a length > 2 cm.	A part of the tail tissue has been bitten off and the tail has been shortened to a length < 2 cm.		
Damage	Nondamaged	Swollen	Bite marks	Wound	Inflamed wound
	No visible damage.	The tail is red and/or swollen. The tail has no bite marks and the skin is not broken.	The tail has bite marks that are seen as small red or black dots on the tail. These can either be seen as bruising without broken skin or as small holes in the skin, but no tissue is missing.	The tail has one or more open wounds. These can vary from scratches (without blood) to a shortened tail with a deep wound (with blood). The wound can have a crust that is intact or has partly fallen off.	The tail is swollen and has one or more open wounds. Can vary from scratches (without blood) to a shortened tail with a deep wound (with blood). The wound can have a crust that is intact or has partly fallen off.
Freshness	No blood	Crust	Red crust	Dark blood	Fresh blood
	No blood on the tail. Can be observed together with an open wound (tail damage score greater than "bite mark") in the case of, for example, small scratches on the tail.	The wound has a crust which is intact or has partly fallen off but where the skin below the crust does not look red or produce any exudate. Can only be observed with an open wound (tail damage score greater than "bite mark").	The wound has a crust which is intact or has partly fallen off and where the skin below the crust looks red but does not produce any exudate. Can only be observed with an open wound (tail damage score greater than "bite mark").	There is dark red/brown blood on the tail which feels wet/sticky (i.e., the wound produces exudate), however touching the wound does not lead to a drop of blood on the finger. Alternatively, when the wound has a partly fallen-off crust with the skin below the crust as described above. Can only be observed with an open wound (tail damage score greater than "bite mark").	Red fresh blood on the tail which feels wet/sticky (i.e., exudate), and where touching the wound leads to a drop of blood on the finger. Alternatively, when the wound has a partly fallen-off crust with the skin below the crust as described above. Can only be observed with an open wound (tail damage score greater than "bite mark").

The tail posture of all pigs was scored weekly on the same day as the lesion scoring. A tail was considered to be curled if the major part of the tail was curled and pointing upwards in relation to the horizontal extension of the back; otherwise the tail was scored as hanging. Tail position was recorded at feeding by filming. Pigs were filmed when all pigs were standing at the feeding trough. The tail position for each pig was then scored by observation of the video recordings. If a pig's tail was not observed during filming, due to the pig sitting down or not visiting the feeding trough, tail position was recorded as missing. Pigs whose tails were so short that the tail position could not be determined were also scored as missing. Filming the entire stable took on average 9.8 min.

Statistical analysis was carried out using the following three software tools, StataIC 15.1 (StataCorp LLC, College Station, TX, USA), MLwiN (Centre for Multilevel Modelning, University of Bristol, Bristol, UK) and SAS 9.4 (SAS Institute, Cary, NC, USA). Data on tail position, tail lesions (length, damage and freshness) and lameness was collected on individual pig level at each sampling occasion. Rows with missing data were removed from the data set. The data in this study was hierarchical with three levels consisting of Pen (highest), Pig and Time (lowest). Each pig was observed at several time points (Time) and hence repeated measurements had to be taken into account.

This was an observational study to investigate the association between tail lesions and tail posture at feeding. A binomial multivariate regression model (trend model) with tail position as the outcome was built in StataIC and MLwiN using the "melogit" command. The statistical unit was pig. Since the recordings of length, damage and freshness are closely correlated (e.g., freshness score equivalent to or more severe than crust was only given to pigs with a damage score of "wound" or "inflamed wound"), only damage was included in the final model. Damage was considered the most suitable choice since it had more variability compared to length and freshness. The analysis included the following independent variables; Pen, Pig, Lame (yes/no), Damage (nondamaged/swollen/bite marks/wound/inflamed wound), Sex (gilt/barrow) and Time (1–14). Time was standardised using the following equation: $sTime = \frac{Time - 7.5}{6.5}$ (Time ranging from 1–14) to receive values of sTime ranging from 0 to 1. The relationship between the data and Time was found not to be linear, and therefore $Time^2$ was used to allow a better linear fit to the data. All variables were included in the model, which was subsequently reduced by backward selection of the significant variables ($p \leq 0.05$). Time and $Time^2$ were kept in the model to account for repeated measurements.

Clustering within pens and pigs was accounted for by including random slopes for sTime at Pen and Pig level and for $sTime^2$ at Pen level. The software was unable to fit a random slope for $sTime^2$ at Pen level. The estimated log likelihood difference between the chosen model and the model with random $sTime^2$ slopes also at Pen level was -18.8614, showing that the model without the random slope at Pen level was better. Furthermore, the analysis of variance showed that there was more variation at pig level than at Pen level, and thus the finalised model without random slope at Pen level was determined to be satisfactory.

The final model was

$$TP_{\pi Time, Pig, Pen} = \beta_{0ID, Pen} + Swollen_{Time, Pig, Pen} + Red_{Time, Pig, Pen} + Wound_{Time, Pig, Pen} + Inflamed\ wound_{Time, Pig, Pen} + Sex(barrow)_{Pig, Pen} + \beta_{5ID, Pen} sTime_{Time, Pig, Pen} + \beta_{6ID, Pen} sTime^2_{Time, Pig, Pen} + e \quad (1)$$

To investigate the possibility of using tail posture as an indicator of tail biting and to estimate the specificity and sensitivity of the method, a receiver operating characteristic (ROC) analysis was conducted between tail posture and tail damage. A curled tail posture was considered to be either present or absent (hanging tail posture). To create the ROC curve, tail position was used to classify tails as Damaged (nondamaged/swollen/bite marks/wound/inflamed wound).

This study comprised only behavioural observations and the clinical scoring of commercial pigs. It was part of a larger trial aimed to improve animal welfare by investigating different ways of providing pigs with straw. Due to the low severity of the treatment the study did not require approval by an ethical committee according to the legislation. All pigs were managed and treated by the staff at the farm according to the ordinary management routines, i.e., injured pigs received appropriate medical treatment, such as the removal of bitten pigs and treatment with antibiotics in the case of severely inflamed wounds or reduced health status.

3. Results

Out of the 5713 observations of pigs scored with curled tails, 55.4% had nondamaged tails, 10.7% had swollen tails, 13.6% had bite marks on the tail, 19.6% had tail wounds and 0.8% had inflamed wounds on the tail. Out of the 379 observations of pigs scored with hanging tails, 28.8% had nondamaged tails, 7.92% had swollen tails, 7.65% had bite marks on the tail, 46.97% had tail wounds and 8.71% had inflamed wounds on the tail.

Both tail posture and tail damage on Pig level varied over time, and the number of damaged tails increased with time. The proportion of hanging tail posture fluctuated, being high in the beginning and in the end of the production period (see Table 1). The proportion of hanging tails and tail damages varied between pigs on Pen level. Two of the observed pens did not have any pigs with hanging tails during the study period, while all other pens had pigs with hanging tails on at least one occasion. Tail damage of some sort was observed in all pens during the study period. The data is provided in Table S1.

3.1. Model

Hanging tails were positively associated with tail damage scores of "wound" and "inflamed wound" ($p < 0.05$), however not with less severe damage, i.e., "swollen" and "bite marks". Pigs with tail damage scored as "wound" were 4.15 times more likely to have hanging tails than pigs with nondamaged tails, while pigs with tail damage scored as "inflamed wounds" were 14.24 times more likely to have hanging tails than pigs with nondamaged tails. Barrows were 1.58 times ($p > 0.046$) more likely to have hanging tails than were gilts (Table 3).

Table 3. Binomial multivariate regression model of hanging tails for 459 pigs in 42 pens over 14 weeks (one observation per week) (n = 6096).

Variables			Odds Ratio (OR)	SE	p-Value
	Fixed variables				
	sTime		1.12	1.3	n.s.
	sTime^2		2.09	1.54	n.s.
Damage (baseline: nondamaged)		Swollen	1.46	1.28	n.s.
		Bite marks	1.21	1.28	n.s.
		Wound	4.15	1.19	<0.0001
		Inflamed wound	14.24	1.48	<0.0001
Sex (baseline gilt)		Barrow	1.58	1.26	0.046
	Constant		0.0059	1.36	
	Random variables				
Pen level		Var (sTime)	1.90	1.34	n.s.
		Var (constant)	1.51	1.26	n.s.
		Cov (sTime, constant)	1.12	1.21	n.s.
Pig level		Var (sTime)	1.40	1.36	n.s.
		Var (sTime^2)	3.30	2.10	n.s.
		Var (constant)	27.41	2.10	n.s.
		Cov (sTime, sTime^2)	1.12	1.39	n.s.
		Cov (sTime, constant)	0.99	1.37	n.s.
		Cov (sTime^2, constant)	0.32	2.03	n.s.

The random slopes for sTime at Pen level and sTime and sTime^2 at Pig level allowed for different Pens and Pigs to develop differences over time. The variation at Pig level was larger than the variation at Pen level (Table 3).

As we found that barrows were more likely to have hanging tails than were gilts, we investigated whether barrows also had more tail damage, as presented in Table 4. When comparing the tail damage of barrows and gilts in a chi-squared test, no differences were found ($p > 0.2688$).

Table 4. Tail damage in gilts and barrows (n = 6125).

Sex	Damage				
	0	1	2	3	4
Gilt	1681	317	391	667	36
Barrow	1604	327	419	634	49

3.2. Evaluation of Tail Posture as an Indicator of Tail Damage

Lowering the cut-off point of what was considered a damaged tail increased the sensitivity (i.e., the probability that a damaged pig tail was classified as damaged through tail position) of tail posture as an indicator of tail lesions (Table 5). Conversely, increasing the cut-off point of what was considered a damaged tail decreased the sensitivity of tail posture as an indicator of tail lesions. Tails scored as "swollen" had a sensitivity of 70.57%, while tails scored as "inflamed wound" had a sensitivity of 8.59%.

Table 5. Receiver operating characteristic (ROC) analysis for quantifying the accuracy of tail posture in identifying tail lesions.

Cut-Off Point	Sensitivity	Specificity	Correctly Classified	Positive Likelihood Ratio	Negative Likelihood Ratio
Nondamaged	100	0	6.28	1	
Swollen	70.57	55.38	56.33	1.5816	0.5314
Bite marks	62.76	66.05	65.84	1.8486	0.5638
Wound	55.21	79.62	78.09	2.7089	0.5626
Inflamed wound	8.59	99.16	93.47	10.2516	0.9218
All	0	100	93.72		1

The specificity (i.e., the probability that an undamaged pig tail was classified as undamaged through tail position) increased when increasing the cut-off point of what was considered a damaged tail. When the cut-off point was set to "inflamed wound", the specificity was 99.16%, while the specificity was 55.38% if the cut-off point was set to "swollen" (Table 5). The percentage of correct classifications increased with an increased cut-off point, increasing from 56.33 to 93.47% as the cut-off point passed from "swollen" to "inflamed wound". The area under the ROC curve, corresponding to the ability to make correct classifications, was 68.62%.

When setting the cut-off point of when to consider a tail as damaged to "wound" or "inflamed wound", the sensitivity was 55.2%, the specificity was 79.7%, and 78.1% were correctly classified. The area under the ROC curve was 67.4%.

4. Discussion

The finding that hanging tails were associated with more severe tail damage, i.e., wounds and inflamed wounds, indicates that a hanging tail posture does not occur until tail damage is already present. This is in line with findings by Larsen et al. [13], showing that pigs with tucked tails were about six times more likely to have a wound the same day. Hence, it does not seem to be enough to look at tail posture in this manner and frequency to identify tail biting behaviour before it has caused tail damage. However, most of the wounds detected by tail posture would not be detectable without close examination of the tail, and therefore tail posture could still be useful in commercial production for detecting tail biting before wounds are detectable from outside the pen.

The correct classification of a damaged tail through its posture was achieved in 67% of cases when the cut-off point was set to "wound", which includes the damages that were statistically proven to be associated with tail posture by our model. This means that 33% of the pigs will be misclassified. When assessing tail posture, 44.8% of the pigs with tail damage will be missed and 20.3% will be misclassified as having wounded tails. This was evident partly from the fact that we observed pigs with hanging tails and undamaged tails (or tails with damage less serious than "wound") while there were also pigs with wounded curly tails (ranging from "swollen" to "inflamed wound"). However, the misclassification of non-injured pigs as injured, which was also reported by Larsen et al. [13], does not have to be considered a large issue. According to a survey of Swedish farmers, the main treatment for stopping tail biting in pigs is the provision of extra straw [17]. Providing extra straw will increase rooting possibilities and could be considered as a positive enrichment while having no negative consequences. However, pigs with wounds that are missed through the assessment of tail position need to be identified through other means such as clinical examination or behavioural observations. The discovery that not only injured pigs have hanging tails is in line with the finding by Lahrmann et al. [15], that in pens without observed tail biting outbreaks approximately 15–17% of pigs had hanging tails, while in pens with observed tail biting outbreaks approximately 23–33% had hanging tails 1–3 days before an outbreak. Similarly, Zonderland et al. [11] found that around 15% of pigs with curled tails had tail damage according to our scale and that 14.6% of the pigs that had tails that were hanging or tucked in between the hind legs were undamaged. Furthermore, Statham et al. [18] also found that there were fewer pigs with hanging tails in groups with no tail biting outbreaks than in groups with tail biting outbreaks. Collectively, this shows evidence that tail posture at feeding could be used as an indicator of tail biting, even though the method has significant flaws and needs further investigation.

Compared to other studies looking at tail posture, (e.g., Zonderland et al. [11], Lahrmann et al. [15], Statham et al. [18]) the pigs in this study were scored less frequently (weekly, compared to daily or three times per week). It is possible that we could have detected tail damages at earlier stages if scoring had been performed more frequently. As found by Lahrmann et al. [15] and Zonderland et al. [11], changes in tail posture can be seen hours to days before tail lesions occur. Due to the experimental setup of our study, we were not able to detect such changes. However, there was no significant effect of time in the statistical model, indicating that time did not have a significant impact on the presence of hanging tails. Moreover, we were unable to associate damage that was less severe than "wound" with tail posture. If we had used shorter observation intervals we may have detected time dependent changes. However, our study was designed to reflect the usability for commercial farmers, who rarely have the time to observe individual pigs as frequently as once per day. We propose that this method could be incorporated into normal farm routines, for instance when farmers are checking feeding equipment or the health status of pigs.

The number of hanging tails was found to fluctuate over time; the frequency was highest at the beginning and the end of the trial period. The beginning of the production period is usually associated with stressful events for the pigs, such as being moved to a new stable, getting new pen mates and receiving new feed or a new feeding regime. Stressful events also occur at the end of the production period, when pigs are larger and the stocking density (kg/m^2) increases. Additionally, larger pigs may be sent to slaughter, which may change the established social hierarchy in the pen, leading to fights. Moreover, some female pigs may become sexually mature, which could also alter behaviour and hierarchy in the pens. As suggested by Kleinbeck and McGlone [12], tail posture might be an indicator of pig comfort. The reason for the large amount of hanging tails observed at the beginning and end of the production period could therefore be related to stress (e.g., due to new environment or new hierarchy), rather than to tail biting. On the other hand, tail biting is also known to occur when stressors such as increased stocking density and new hierarchies arise [2]; however, the causality for hanging tails is not evident from this study.

Lahrmann et al. [15] noted that the activity of pigs influenced their tail posture, revealing that tail posture could also be a response to emotional states other than, e.g., pain or discomfort. For example, pigs engaged in rooting activities were more likely to have hanging tails compared to walking or running pigs. Furthermore, tail posture was more likely to change in a short time after pigs changed activities. As discussed by de Oliveira and Keeling [19], certain animal postures may not be specific to specific emotional states and may not be possible to assess alone but rather only in combination with whole body posture. They found that there were interactions between tail, ear and neck position and activity in cows, suggesting that cows express themselves differently during different activities (brushing, queuing for milking or feeding). In the present study we scored the tail posture at feeding. There is a lack of knowledge about how feeding alone is associated with the tail posture of pigs. From the results of this study we know that the majority of the pigs had curled tails at feeding. However, some of the curled tails had severe damage. It is possible that feeding alone makes pigs curl their tails; for example, since no other pigs are situated behind the pigs at this time, there should be no need to protect the tail between the hind legs. To completely understand the association between tail position and tail damage, the tail position should be investigated further during several different activities and emotional states. When designing the present study, the decision to score the tail posture at feeding was made for two main reasons. Firstly, it is commonly observed that most pigs have curled tails while eating. Secondly, the scoring of tail posture, especially in commercial settings, would be made easier if all pigs in a pen could be rapidly scored. When pigs in long trough pens are fed, the feeding system releases feed in one pen at the time. The next pen in line will receive feed a few seconds after the previous pen, making it possible for us to easily score tail posture on the feeding pigs. This would make it possible for farmers to follow the feeding loop and observe one pen (commonly with ~10 pigs per pen in Sweden) at a time. This assessment could then be easily incorporated into the management scheme. In this study we recorded the pigs at feeding by video in order to score the tail posture afterwards. This was performed mainly to register the ID of each pig and subsequently combine tail posture with tail scoring. In commercial farms, it would be possible merely to note in which pens there are numerous pigs with hanging tails to obtain an indication of where there could be tail biting issues. Hence, these direct observations could ease the implementation of this technique in common management practices. The time taken to observe all pens (42 pens, 459 pigs) in this trial was on average less than ten minutes at experimental setting (which is considered to be more time consuming). This suggests that this method is also feasible in systems where there are more than 10 pigs per pen as long as the animals all feed at the same time and the feeding trough is located so that the pigs eat standing next to each other, thus facilitating observation for the pig keeper.

The random slopes in the developed model took into account that different pigs may react differently. The variability was higher at individual than at pen level, suggesting that the mix of pen mates might even out individual differences. It is known that pain is a subjective experience and may cause different reactions to the same stimuli (e.g., Ison et al. [20]). Therefore, some pigs may react with a hanging tail when another pig is manipulating their tail, while other pigs might not react at all even when there is a wound on the tail. Different pigs may also have different coping strategies for avoiding tail biting. As hypothesised by Feddes and Fraser [21], a curled tail could also be a way of reducing tail biting by protecting the tip of the tail, which is the most commonly attacked part. This is converse to the hypothesis that the pig uses a hanging tail posture in order to protect the tail [12–15].

The pigs in this study were either provided with straw through a rack or directly on the floor, as part of another experiment. It is well known that straw provision has an impact on the development of tail biting (e.g., Schroder-Petersen et al. [2], Wallgren et al. [22]). It is therefore likely that the straw provision affects the occurrence of tail biting, however the association between tail lesions and tail posture is likely not affected. The aim of this study was mainly to investigate the association between tail position and tail damage, which we do not consider to be substantially affected by straw provision.

Larsen et al. [13] found a higher probability of having a lowered tail in pigs that were not provided with straw compared to pigs provided with straw; however, the underlying cause for this is not fully understood, and could perhaps be an indicator of the impact of the environment on the emotional state of the pig. As previously discussed, more research is needed to further investigate tail posture in relation to the emotional state of animals [2,12]. Tail biting is multifactorial and has been known also to be affected by, e.g., feeding regime [2]. During this trial, the feeding was altered from four to three meals per day due to the farm normal feeding regimes. This may have affected the occurrence of tail biting. However, all pens experienced the same changes in feeding at the same time, and therefore all pens are likely affected in the same way, and the association between tail lesions and tail posture are assumed to not likely be affected.

The causal relationship between tail posture and tail biting/tail damage is not fully understood. From our study, where tail posture and tail damage was only scored once per week and there was no significant effect of time, it is not possible to discriminate which occurred first, the tail damage or the hanging tail. The causality is therefore unknown. However, Zonderland et al. [11] and Lahrmann et al. [15] found that in weaned piglets hanging tails occurred days before tail damage could be detected. This implies that hanging tails could be an indicator that pigs are trying to protect the tail, although the protection often seems to fail since damage occurs anyway. However, we cannot rule out the possibility that hanging tails are simply more prone to being bitten. According to Kleinbeck and McGlone [12], the tails of pigs that that were being bitten were in the hanging posture. However, as previously mentioned, as observed in both this and other studies, hanging tails are not always damaged [11].

5. Conclusions

A hanging tail posture at feeding was found to be significantly correlated to wounds and inflamed wounds on pig tails. Pigs with tail wounds were four times more likely to have hanging tails compared to pigs with undamaged tails. Pigs with inflamed tail wounds were 14 times more likely to have hanging tails compared to pigs with undamaged tails. Only by considering tail position at feeding, 78% of the pigs were able to be correctly classified as having tail wounds or inflamed tail wounds or not. Even though the tail position at feeding is not fully accurate in identifying animals with bitten tails, it is considered feasible in commercial circumstances.

Supplementary Materials: The following are available online at http://www.mdpi.com/2076-2615/9/1/18/s1: Table S1: Data—kopia.

Author Contributions: Conceptualisation, T.W., A.L. and S.G.; Investigation, T.W. and A.L.; Data Curation, T.W.; Formal Analysis, T.W.; Writing—Original Draft Preparation, T.W.; Review and Editing, A.L. and S.G.; Supervision, S.G.; Funding Acquisition, S.G.

Funding: This research was funded by ERA-Net ANIHWA; FareWellDock, The Swedish Research Council Formas, grant number 221-2013-1799, and the Faculty of Veterinary Medicine and Animal Sciences at the Swedish University of Agricultural Sciences. Materials and equipment were funded by "Djurvännernas förening. Stockholm" [Friends of animals association in Stockholm].

Acknowledgments: We would like to thank the producer for participating in the experiment.

Conflicts of Interest: The authors declare no conflicts of interest.

References

1. European Food Safety Authority (EFSA). Scientific Opinion of the Panel on Animal Health and Welfare on a request from Commission on the risks associated with tail biting in pigs and possible means to reduce the need for tail docking considering the different housing and husbandry systems. *EFSA J.* **2007**, *611*, 1–13.
2. Schroder-Petersen, D.L.; Simonsen, H.B. Tail biting in pigs. *Vet. J.* **2001**, *162*, 196–210. [CrossRef] [PubMed]
3. *Council Directive 2008/120/EC Laying Down Minimum Standards for the Protection of Pigs*; 2008/120/EC, C.D.; The Council of the European Union: Brussels, Belgium, 2008.

4. D'Eath, R.B.; Niemi, J.K.; Ahmadi, B.V.; Rutherford, K.M.D.; Ison, S.H.; Turner, S.P.; Anker, H.T.; Jensen, T.; Busch, M.E.; Jensen, K.K.; et al. Why are most EU pigs tail docked? Economic and ethical analysis of four pig housing and management scenarios in the light of EU legislation and animal welfare outcomes. *Animal* **2016**, *10*, 687–699. [CrossRef] [PubMed]
5. Lerner, H.; Algers, B. Tail docking in the EU: A case of routine violation of an EU Directive. In *The Ethics of Consumption*; Röcklinsberg, H., Sandin, P., Eds.; Wageningen Academic Publishers: Wageningen, The Netherlands, 2013.
6. Lahrmann, H.P.; Busch, M.E.; D'Eath, R.B.; Forkman, B.; Hansen, C.F. More tail lesions among undocked than tail docked pigs in a conventional herd. *Animal* **2017**, *11*, 1825–1831. [CrossRef] [PubMed]
7. Bolhuis, J.E.; Schouten, W.G.P.; Schrama, J.W.; Wiegant, V.M. Behavioural development of pigs with different coping characteristics in barren and substrate-enriched housing conditions. *Appl. Anim. Behav. Sci.* **2005**, *93*, 213–228. [CrossRef]
8. Taylor, N.R.; Main, D.C.J.; Mendl, M.; Edwards, S.A. Tail-biting A new perspective. *Vet. J.* **2010**, *186*, 137–147. [CrossRef] [PubMed]
9. D'Eath, R.B.; Arnott, G.; Turner, S.P.; Jensen, T.; Lahrmann, H.P.; Busch, M.E.; Niemi, J.K.; Lawrence, A.B.; Sandoe, P. Injurious tail biting in pigs: How can it be controlled in existing systems without tail docking? *Animal* **2014**, *8*, 1479–1497. [CrossRef] [PubMed]
10. Fraser, D. Attraction to Blood as a Factor in Tail-Biting by Pigs. *Appl. Anim. Behav. Sci.* **1987**, *17*, 61–68. [CrossRef]
11. Zonderland, J.J.; van Riel, J.W.; Bracke, M.B.M.; Kemp, B.; den Hartog, L.A.; Spoolder, H.A.M. Tail posture predicts tail damage among weaned piglets. *Appl. Anim. Behav. Sci.* **2009**, *121*, 165–170. [CrossRef]
12. Kleinbeck, S.; McGlone, J.J. *Pig Tail Posture: A Measure of Stress*; Texas Tech University Agricultural Science Technical Report No.-T-327; Texas Tech University: Lubbock, TX, USA, 1993; pp. 47–48.
13. Larsen, M.L.V.; Andersen, H.M.L.; Pedersen, L.J. Tail posture as a detector of tail damage and an early detector of tail biting in finishing pigs. *Appl. Anim. Behav. Sci.* **2018**. [CrossRef]
14. McGlone, J.J.; Sells, J.; Harri, S.; Hurst, R.J. *Cannibalism in Growing Pigs: Effects of Tail Docking and Housing System on Behaviour, Performance and Immune Function*; Texas Tech University Agricultural Science Technical Report No. T-5-283; 1990; Volume 283, pp. 69–71.
15. Lahrmann, H.P.; Hansen, C.F.; D'Eath, R.; Busch, M.E.; Forkman, B. Tail posture predicts tail biting outbreaks at pen level in weaner pigs. *Appl. Anim. Behav. Sci.* **2018**, *200*, 29–35. [CrossRef]
16. Zonderland, J.J.; Fillerup, M.; Reenen, C.G.; Hopster, H.; Spoolder, H.A.M. *Preventie en Behandeling van Staartbiljten Bij Gespeende Biggen*; Wageningen UR: Animal Sciences Group: Lelystad, The Netherlands, 2003.
17. Wallgren, T.; Westin, R.; Gunnarsson, S. A survey of straw use and tail biting in Swedish pig farms rearing undocked pigs. *Acta Vet. Scand.* **2016**, *58*, 11. [CrossRef] [PubMed]
18. Statham, P.; Green, L.; Bichard, M.; Mendl, M. Predicting tail-biting from behaviour of pigs prior to outbreaks. *Appl. Anim. Behav. Sci.* **2009**, *121*, 157–164. [CrossRef]
19. De Oliveira, D.; Keeling, L.J. Routine activities and emotion in the life of dairy cows: Integrating body language into an affective state framework. *PLoS ONE* **2018**, *13*, e0195674. [CrossRef] [PubMed]
20. Ison, S.H.; Clutton, E.R.; Di Giminani, P.; Rutherford, K.M.D. A Review of Pain Assessment in Pigs. *Front. Vet. Sci.* **2016**, *3*, 108. [CrossRef] [PubMed]
21. Feddes, J.J.R.; Fraser, D. Non-nutritive chewing by pigs—Implications for tail-biting and behaviourial enrichment. *Trans. ASAE* **1993**, *93*, 521–527.
22. Wallgren, T.; Larsen, A.; Lundeheim, N.; Westin, R.; Gunnarsson, S. Implication and impact of straw provision on behaviour, lesions and pen hygiene on commercial farms rearing undocked pigs. *Appl. Anim. Behav. Sci.* **2018**. [CrossRef]

© 2019 by the authors. Licensee MDPI, Basel, Switzerland. This article is an open access article distributed under the terms and conditions of the Creative Commons Attribution (CC BY) license (http://creativecommons.org/licenses/by/4.0/).

Article

Are Tail and Ear Movements Indicators of Emotions in Tail-Docked Pigs in Response to Environmental Enrichment?

Míriam Marcet-Rius [1,*], Emma Fàbrega [2], Alessandro Cozzi [3], Cécile Bienboire-Frosini [1], Estelle Descout [4], Antonio Velarde [2] and Patrick Pageat [5]

1. Physiological and Behavioral Mechanisms of Adaptation Department, IRSEA (Research Institute in Semiochemistry and Applied Ethology), Quartier Salignan, 84400 Apt, France
2. Animal Welfare Program, IRTA (Insitute of Agrifood Research and Technology), Centre de Control i Avaluació de Porcí, 17121 Monells, Girona, Spain
3. Research and Education Board, IRSEA (Research Institute in Semiochemistry and Applied Ethology), Quartier Salignan, 84400 Apt, France
4. Statistical Analysis Service, IRSEA (Research Institute in Semiochemistry and Applied Ethology), Quartier Salignan, 84400 Apt, France
5. Semiochemicals' Identification and Analogs' Design Department, IRSEA (Research Institute in Semiochemistry and Applied Ethology), Quartier Salignan, 84400 Apt, France
* Correspondence: m.marcet@group-irsea.com

Received: 26 April 2019; Accepted: 9 July 2019; Published: 16 July 2019

Simple Summary: The assessment of animal welfare should involve physical and mental welfare. Physical welfare is relatively easy to measure. Nevertheless, there is a lack of feasible indicators of positive emotions in farm animals, which causes difficulties in obtaining a complete analysis of welfare. To improve the quality of life of farm animals, it is necessary to be able to assess their welfare in a valid and feasible way. This study aimed to determine whether environmental enrichment, understood to be positive for animal welfare, could influence tail movement and ear movement in fattening pigs. These indicators could be used to assess emotions, with positive or negative valences, as suggested in previous studies on mini-pigs and pigs. The results showed that tail movement was a valid and feasible indicator of positive emotions; pigs moved their tails a greater number of times when they interacted more with enrichment than when they interacted less with it. Regarding ear movements, this study revealed a need for further investigation. This research could play an important role in improving the analysis of different emotions in pigs, thereby improving the assessment of animal welfare in pig breeding systems using valid, feasible, and noninvasive indicators of emotions.

Abstract: The inclusion of emotional indicators in farm monitoring methods can improve welfare assessments. Studies in controlled conditions have suggested that increased tail movement is an indicator of positive emotions in pigs, while others have proposed that increased ear movements are linked to negative emotions. This study aimed to investigate these indicators in pig farm conditions to analyze their validity and the effect of enrichment on welfare. Thirty-six pigs received one of the following enrichment materials: straw in a rack, wooden logs, or chains. Behavioral observations were performed by focal sampling. The results showed that tail movement duration was significantly higher when pigs exhibited "high use" (three or more pigs in a pen interacting with the enrichment) than when they exhibited "low use" (fewer than three) of enrichment ($p = 0.04$). A positive correlation was found between tail movement frequency and duration ($r = 0.88$; $p = 0.02$). The increase in tail movement could be considered an indicator of positive emotions in pigs when measured with other categories of indicators. Regarding ear movements, no significant difference was found. Future studies should further investigate these indicators thoroughly, as the results could be useful for improving the assessment of emotions in pigs.

Keywords: animal welfare; pig assessment; positive emotions; negative emotions; enrichment material

1. Introduction

The assessment of farm animal welfare requires a good understanding of the animals' affective experiences, including their emotions [1]. Emotions are transient reactions to short-term triggering events, and their continued occurrence can cause longer-lasting affective states, which represent good or poor welfare [2]. The inclusion of indicators of emotions in farm monitoring methods can improve welfare assessments beyond the traditional focus on the mere absence of disease and distress [3] and good feeding and housing practices [4].

Currently, animal welfare assessments of farm animals do not always include the assessment of emotions, either positive or negative. In addition, when they are included, they can be influenced by a subjective assessment of the auditor, for example, in the case of the Qualitative Behaviour Assessment [5]. Thus, more objective and feasible indicators of the emotions of farm animals could be very useful for providing new insights into both animal welfare assessments and our understanding of the welfare state, either positive or negative, of farm animals [3].

Emotional experiences are valenced, being perceived as positive or negative, rewarding or punishing, pleasant or unpleasant [6]. Emotional experiences also vary in terms of reported activation or arousal [6]. Emotional arousal can be defined as an emotional activation, in which animals' bodies experience heightened physiological activity, and extremes of emotion, being positive, such as excitement, or negative, such as anger [7].

Some authors [8,9] have suggested that an increase in tail movement for pigs is related to positive emotions: both tail wagging and changes in tail posture occur more often during rewarding events than during aversive events. Other studies have also reported that increased tail wagging in pigs is related to positive situations such as social greetings [10,11] and play [12]. Recent studies performed under controlled conditions with mini-pigs provided clear results: tail movement duration was significantly higher when animals performed play behaviors when provided with enrichment materials (medium-sized dog toys) than when animals did not play because they were not provided with enrichment materials [13,14]. Play is believed to trigger positive emotions [15,16], which suggests that a long tail movement duration reflects positive emotions in pigs.

In contrast, Reimert and colleagues [8] suggested that increased frequency of ear posture changes tended to be linked to negative situations and were less frequent in positive situations. Marcet-Rius et al. [14] showed that under controlled conditions, ear movement frequency was significantly lower in a group of mini-pigs that was allowed to play with an object than in a group that was not allowed to do so. Thus, the study provided new information about this potential indicator of emotions in pigs by demonstrating that pigs showed fewer ear movements in a positive situation than in a control situation (with no stimulus or enrichment). Another study performed under controlled conditions with mini-pigs showed that ear movement frequency and the frequency and duration of other known indicators of poor welfare (agonistic behavior and displacement behavior) were significantly lower in a group provided with straw than in a control group that was not provided with any manipulatable material or other enrichment [17]. The provision of straw in pig production systems is widely presumed to be beneficial to animal welfare [18,19], and the observation that ear movement frequency was significantly lower in pigs provided with straw than that during the control session, together with two other indicators of poor welfare, strongly suggests that a high frequency of ear movement is more likely to be linked to negative emotions or at least is more common in a poor environment. Similar results regarding ear posture changes have been found in other species, such as sheep [20–22], dogs [10], and horses [23].

Intensive production systems are often very barren with concrete (slatted) floors and no substrate in which the animals can root [24]. Such environments hamper the ability of pigs to express some key behaviors, such as exploration and foraging [25]. As a consequence, harmful and manipulative behaviors such as ear and tail biting often occur at high frequencies [24]. Successful enrichment should decrease the incidence of abnormal patterns of behavior and increase the performance of behaviors such as exploration, foraging, play, and social interaction, which are within the range of the animal's normal, species-specific behavior [25]. Additionally, enrichment could also enhance performance by improving, for example, the feed conversion ratio [26]. Recent studies have shown that pigs have more optimistic judgment biases in enrichment environments, a fact that indicates a more positive affective state and, hence, better welfare [27]. The provision of appropriate environmental enrichment to pigs is mandatory by law in Europe [28]. Additionally, the European Union encourages leaving pigs undocked per the EU Recommendation of 2016 [29]. Tail-docking is a procedure consisting of cutting the tail of pigs, sometimes without anesthesia and analgesia, which has been commonly used to reduce the risk of tail biting. Tail-biting incidents also occur when tails are docked; therefore, docking as such does not solve the tail-biting problem [29]. Thus, the need to identify indicators of positive welfare that could support the use of enrichment materials to reduce the incidence of tail biting is important. Tail movement could be a practical tool for farmers to assess positive animal welfare.

Assuming that increases in tail movement and ear movement are respective indicators of positive and negative emotions in mini-pigs in a controlled system, the aim of the present study was to investigate these potential indicators in pigs provided with enrichment materials at an experimental farm. We hypothesize that increases in tail and ear movements could be indicators of positive and negative emotions in pigs, respectively. In this study, enrichment materials were provided to pigs to allow exploratory behavior, which is very important in these animals [30]. We measured how the interaction with the enrichment material influenced these potential indicators of emotions in pigs.

2. Materials and Methods

The housing, husbandry and use of animals for the procedures described in this manuscript were carried out according to Spanish and European legislation. The project, including this experimental procedure, was approved by IRTA's (Institute of Agrifood Research and Technology, Caldes de Montbui, Spain) Ethics Committee (approval number: AGL2015-68373-C2-2-R).

2.1. Animals and Housing

The pigs (*Sus scrofa domesticus*) (n = 36; entire males) involved in this study were a conventional commercial cross between Landrace x Large White dams with a Pietrain sire. All the pigs came from the same farm (Batallé® selecció, Riudarenes, Girona, Spain), and they arrived at the experimental farm at IRTA at two months of age. All of the pigs were tail-docked: the commercial farm of origin was in the process of implementing an action plan (i.e., undertaking correction actions to prevent tail biting if necessary) after carrying out a risk analysis of tail biting as required by the EU Recommendation of 2016 [29]; they will start leaving pigs undocked when the action plan is proven to work, first with docked pigs. Additionally, the results of a preliminary trial conducted on the same research farm by the authors revealed that using some of the single enrichment items used in this study, under similar conditions, led to an increased occurrence of tail biting. Therefore, due to ethical concerns, the authors decided to perform the present study in docked pigs as a first step towards moving to undocked pigs. The pigs entered the present study at 2.5 months of age. Thirty-six pigs were involved in the study and were divided into two identical rooms with 18 pigs per room. In each room, there were three identical pens with six animals in each pen, with a stocking density of 0.9 m^2/pig. In every pen, each pig had a different colored tag (blue, yellow, red, orange, green, and white) to differentiate it from other pigs. Three different types of enrichment materials were constantly provided to the animals for three months, once for each pen in each room: straw in a rack, wooden logs, or chains. More precisely, there was one pen in each room with each type of enrichment material, totaling two pens for each material (12 pigs for each material, divided into two pens of 6 animals) (Figure 1).

Room 1	Room 2
Pen 1: 6 pigs	Pen 4: 6 pigs
Enrichment: straw in a rack	Enrichment: wooden logs
Pen 2: 6 pigs	Pen 5: 6 pigs
Enrichment: wooden logs	Enrichment: chains
Pen 3: 6 pigs	Pen 6: 6 pigs
Enrichment: chains	Enrichment: straw in a rack

Figure 1. Distribution of different types of enrichment material.

For this experimental procedure, the most important analysis was the interaction or lack of interaction with the enrichment, regardless of type, as a way to create a positive situation for the animals and a control situation (lack of interaction). All the animals were provided enrichment materials, meaning that all of them could interact with the materials and that no animals were in a nonenriched pen. All the pigs were maintained under the same housing conditions and managed in the same way and by the same stock people. The rooms where the pigs were housed and where the experiment was carried out had an automatic control system for regulating temperature (22 ± 5 °C) and ventilation. The pens had a completely slatted floor. The pigs were fed ad libitum with a commercial pig diet. More precisely, a commercial concentrate was provided in a phased feeding regime (15.04% crude protein and 2.321 kcal net energy at mid-fattening). Animals had continuous access to drinking water via bowl-type drinkers. The pigs in our study were part of a broader investigation in which other behavioral and physiological indicators related to exploratory behavior were collected. Our study on tail and ear movements ended at 5.5 months of age, and the whole investigation ended one week later, when the pigs were 170 days of age. Concerning the period of the year, the present study started at the beginning of March and finished at the end of May. Pigs were slaughtered in a commercial abattoir after experiencing two different transport conditions for the mentioned broader investigation [31].

2.2. Procedure

Thirty-six pigs received one type of environmental enrichment material: twelve pigs were exposed to one of the three types of enrichment material, either straw in a rack, wooden logs or chains. Straw (nonchopped) was continuously provided in a rack ($60 \times 40 \times 80$ cm) and not on the floor to avoid large amounts of straw going underneath the slat system and to reduce costs for the farmers. Two wooden logs were hung horizontally, perpendicularly from a chain in each pen (one on the door and another beside the door). The wooden logs ($30 \times 5 \times 5$ cm) were near the floor but not touching it. They were made from fresh wood from an ash tree and were replaced after 20 days of use. Two chains (50 cm long) were attached vertically, perpendicularly to the metal fence bars of each pen (one on the door and another beside the door), and as with the wooden logs, they were near the floor but not touching it. Chains are considered of marginal interest according to the EU recommendation [29]. Materials of marginal interest can provide distraction but should not be considered to fulfill the essential needs of the pigs [29]. Thus, such materials do not elicit sufficient exploratory behavior, but this does not mean this type of enrichment is negative for the pigs and that pigs do not interact with it. Additionally, this type of enrichment, despite its low interest, is still widely used on farms. For this reason, this type of enrichment was used for the present study, the aim of which was to investigate tail and ear movements as potential indicators of emotions when pigs were interacting or not interacting with the enrichment, even if the authors acknowledge that the effect of chains may be less positive than that of other enrichment materials. Over three months, behavioral observations were performed once per week by one observer and with the help of one technician: the observer indicated the beginning and ending of each behavior (tail movement and ear movements), and the technician noted the frequencies and durations of each behavior with the help of a stopwatch on a data collection sheet. The interaction with the enrichment materials was scored as a "yes" or "no" for each individual during the two-minute observation. More specifically, focal samplings of each pen were performed every Wednesday morning for 12 weeks, from 9 a.m. to 1 p.m., using a direct observation data collection sheet. The behaviors were described as follows: tail movement, ear movement and interaction with the enrichment. Tail movement was defined as tail swinging in any direction, but mostly from side to side (lateral tail movements) [9,10]; ear movement could be defined as any ear movement or ear posture change, including one or two ears (i.e., changes between 'front' and 'back ear postures') [9,14,17]; and interaction with the enrichment was defined as any manipulation, exploration, and snout contact with the enrichment material. The frequency and duration of tail movement were analyzed. Tail movement frequency indicates the number of times that a pig starts moving the tail from side to side during the two-minute period. Tail movement duration is expressed as a percentage and means that a pig is moving its tail, and a new movement is considered to begin when it stops the movement for at least two seconds [13,14]. Ear movement is measured only as a frequency because it is considered an event, characterized by behavior patterns of such a short duration that they are difficult to measure over time [32]. Behaviors could overlap and were not considered mutually exclusive. Each animal was observed for two minutes. Before starting the observations, the operators waited for 2 min in front of each pen to allow the animals to acclimate to their presence to minimize any influence on their behavior.

2.3. Statistical Analysis

Data analysis was performed using SAS 9.4 software (copyright (c) 2002–2012 by SAS Institute Inc., Cary, NC, USA). The significance threshold was fixed at 5%. The experimental unit was the pen: thus, N = 6. First, differences between the three types of enrichment materials in relation to the three variables (tail movement frequency and duration and ear movement frequency) were analyzed over all 12 weeks. Before analysis, the assumption of the normality of the model residuals was verified with the UNIVARIATE procedure using residual diagnostics plots; homogeneity of variances of the data was verified with the GLM procedure using Levene's test (using the HOVTEST = LEVENE option in the MEANS statement). As these conditions were met, a General Linear Mixed Model (including the room as a random effect) was carried out using the MIXED PROCEDURE, with multiple comparisons being performed using the LSMEANS statement in the MIXED procedure with the option ADJUST = TUKEY.

The second part of the statistical analysis consisted of a comparison of the variables (tail movement frequency, tail movement duration and ear movement frequency) over all 12 weeks in relation to the use of enrichment regardless of type, which was scored as "high" or "low": high use of enrichment was considered when three or more animals in the pen interacted with the enrichment material during the 2-min observation of each animal, for a total of six animals in the pen, while low use was considered when fewer than three pigs interacted with the enrichment material during the 2-min observation of each animal. Before analysis, the assumption of the normality of the model residuals was verified as mentioned before. As these conditions were met, a General Linear Mixed Model (including the room as a random effect) was carried out using the MIXED PROCEDURE, with multiple comparisons being performed using the LSMEANS statement in the MIXED procedure with the option ADJUST = TUKEY. See the Supplementary Materials for a detailed description of the full statistical model.

The third part of the analysis consisted of the correlations among the three variables to understand their potential relationships. As the normality of model residuals was verified (UNIVARIATE procedure), correlations among the three variables (tail movement frequency, tail movement duration, and ear movement frequency) for the sum of the 12 weeks were assessed with Pearson's correlation coefficient using the CORR procedure. According to Martin and Bateson [32], r = 0.4–0.7 is considered to indicate a moderate correlation (substantial relationship), r = 0.7–0.9, a high correlation (marked relationship) and r = 0.9–1.0 a very high correlation (very dependable relationship).

3. Results

3.1. Comparison of the Variables (Tail Movement Frequency, Tail Movement Duration and Ear Movement Frequency) over all 12 Weeks in Relation to the Type of Enrichment Material

The aim of the study was to investigate potential indicators of emotions (tail movement frequency, tail movement duration and ear movement frequency) in pigs provided with enrichment material at an experimental farm. First, it was necessary to investigate if the different types of enrichment could affect our variables. No significant differences were found among the three types of enrichment materials with regard to tail movement frequency (mean values for enrichment over all 12 weeks: straw rack = 2.19; wooden logs = 2.25; and chains = 2.08; df = 2; F = 0.33; p = 0.72), tail movement duration (%) (mean values for enrichment over all 12 weeks: straw rack = 30.67; wooden logs = 28.21; chains = 27.95; df = 2; F = 0.36; p = 0.70); or ear movement frequency (mean values for enrichment over all 12 weeks: straw rack = 1.29; wooden logs = 1.38; and chains = 1.40; df = 2; F = 0.11; p = 0.90) (Table 1). Therefore, the type of enrichment was not considered in subsequent analyses, and we focused on only whether enrichment materials were being used by the pigs.

Table 1. Comparison of the variables (tail movement frequency, tail movement duration and ear movement frequency) over all 12 weeks in relation to the type of enrichment material.

Variable		Type of Enrichment			df	F	p-Value
		Straw Rack	Wooden Logs	Chains			
TMF	N	24	24	24	2	0.33	0.72
	Minimum	0.8	0.3	0.0			
	Maximum	4.0	3.8	3.7			
	Mean	2.3	2.3	2.1			
	SE	0.2	0.2	0.2			
TMD (%)	N	24	24	24	2	0.36	0.70
	Minimum	3.2	1.5	0.0			
	Maximum	58.5	52.9	59.9			
	Mean	30.7	28.2	28.0			
	SE	2.9	2.5	3.0			
EMF	N	24	24	24	2	0.11	0.90
	Minimum	0.0	0.0	0.5			
	Maximum	3.3	4.0	3.8			
	Mean	1.3	1.4	1.4			
	SE	0.2	0.2	0.2			

TMF: Tail movement frequency; TMD: Tail movement duration (%); EMF: Ear movement frequency; N: Number of pens (2 pens for each type of enrichment for 12 weeks); SE: Standard error.

3.2. Comparison of the Variables (Tail Movement Frequency, Tail Movement Duration and Ear Movement Frequency) over all 12 aweeks in Relation to the Use of Enrichment (Scored as "High" or "Low") Regardless of Type

A trend, even if not statistically different, was found between the pens with a low or high use of enrichment material for tail movement frequency, which was higher for a high use of enrichment (mean for high use = 2.50; mean for low use = 1.89; df = 1; F = 3.76; p = 0.06). A statistically significant difference was found for tail movement duration (%), which was higher for high use of enrichment (mean for high use = 33.15; mean for low use = 25.17; df = 1; F = 4.88; p = 0.04). No statistically significant difference was found for ear movement frequency (mean for high use = 1.25; mean for low use = 1.45; df = 1; F = 0.28; p = 0.60) (Table 2).

Table 2. Comparison of the variables (tail movement frequency, tail movement duration and ear movement frequency) over 12 weeks in relation to the use of enrichment (scored as "high" or "low").

Variable		Use EM		df	F	p-Value
		High	Low			
TMF	N	34	38	1	3.76	0.06
	Minimum	0.5	0.0			
	Maximum	3.8	4.0			
	Median	2.7	1.8			
	SE	0.1	0.2			
TMD (%)	N	34	38	1	4.88	0.04
	Minimum	1.5	0.0			
	Maximum	59.9	54.4			
	Median	32.4	25.3			
	SE	13.4	12.6			
EMF	N	34	38	1	0.28	0.60
	Minimum	0.0	0.0			
	Maximum	4.0	3.3			
	Median	1.1	1.3			
	SE	0.2	0.1			

Use EM: Use of enrichment material; High use of enrichment: when three or more pigs in a pen interact with the enrichment material; Low use of enrichment: when fewer than three pigs in a pen interact with the enrichment material; TMF: Tail movement frequency; TMD: Tail movement duration (%); EMF: Ear movement frequency; N High: Number of pens in which the use of enrichment was scored as high; N Low: Number of pens in which the use of enrichment was scored as low; N total = 6 pens × 12 weeks = 72.

3.3. Correlations Among the Three Variables (Tail Movement Frequency, Tail Movement Duration and Ear Movement Frequency) for Data Summed over the 12 Weeks

A positive correlation was found between tail movement frequency and tail movement duration (r = 0.88; p = 0.02). No other relevant correlations were found between tail movement frequency and ear movement frequency (r = 0.42; p = 0.41) nor between tail movement duration and ear movement frequency (r = 0.03; p = 0.95).

4. Discussion

The results showed that tail movement duration was significantly higher when the animals interacted more with the enrichment materials than when they with it interacted less during the fattening period. A trend, even if not statistically different, was also found for tail movement frequency, which was also higher with a high use of enrichment. A positive correlation was found between tail movement duration and tail movement frequency. This finding suggests that a high tail movement duration could be an indicator of emotions in fattening pigs with a positive outcome, according to the literature [8,9,13,14,33]. Furthermore, it suggests that a high tail movement frequency could be linked to positive emotions, even if more studies are needed for confirmation, taking into account different total durations of observation and perhaps the ratio between these two parameters. These results suggest that tail movement in pigs could be linked to the use of enrichment materials and therefore to exploratory behavior, which is very important in pigs. The tail movement could also be used to indicate positive emotions, which would indicate positive animal welfare, together with other categories of indicators. Nevertheless, more studies about enrichment materials and different indicators of emotions could be very useful to investigate thoroughly these hypotheses: (i) that enrichment material produces positive emotions to pigs and (ii) that an increase of tail movement indicates positive emotions.

Concerning the relationships among the three parameters (tail movement frequency, tail movement duration and ear movement frequency), the results showed a positive correlation between tail movement frequency and tail movement duration. This finding suggests that an increase in tail movement, either in frequency or duration, could be linked to positive emotions and that both could be used as indicators of emotions with a positive outcome. Previous studies [14] have suggested that, over a ten-minute observation period, a high tail movement duration could be a useful indicator, although the results for frequency were not significant. These results suggest that tail movement could be a valid indicator to assess emotions, together with other categories of indicators. Apart from the duration of tail movement during a set period of time, another feasible measure may be the mean duration of tail movement episodes expressed as a ratio of duration to frequency, but more research is necessary to confirm this notion.

Interestingly, the results of this study also indicated that mini-pigs could be a suitable model of domestic commercial pigs, at least for the parameters observed in the present study, since the results were in accordance with previous studies with mini-pigs. Some studies have examined behavior and welfare in different breeds of mini-pigs [34], and other studies have suggested that welfare indicators of commercial pigs can be used for mini-pigs [35]. Nevertheless, until now, no studies have shown that mini-pigs seem to be a suitable model of domestic commercial pigs, at least for these behavioral parameters.

It is important to remember that all the animals in this study received some type of enrichment material, meaning that all of them could use these materials and that none of the animals were in a nonenriched pen. Tail biting was not observed. The absence of a control group represents a limitation of our study as we had to compare pens only in terms of using more or less enrichment material. Providing enrichment materials to all pigs is mandated by law [28], and many studies have already shown their benefits [19,30,36]. Therefore, in farm conditions, which were used in the present study, no breeders should house their pigs without enrichment materials. The lack of a control group turned into a positive aspect; even though all the animals had the possibility of using enrichment materials, and thus were not in a completely barren environment, they still showed more tail movement when they interacted with the enrichment materials than when they did not. This result suggests that the pigs were more aroused and demonstrated a positive valence [6] during this interaction and that it triggered higher expression of the response behavior (tail movement). In general, the results showed a longer tail movement duration when the animals used the enrichment materials or used them more frequently than when they did not use them or used them less frequently, which suggests that tail movement is linked to positive emotions and therefore to animal welfare. These results are in accordance with the findings of previous studies [9,10,14]. Previous work [13,14] has shown similar results in mini-pigs in an experimental system, and the present study shows that these results can be reproduced at large experimental farms.

Concerning the different types of enrichment material used for the study, although the aim was not to investigate what enrichment was more adequate for the animals, but only if the enrichment, regardless the type, could increase tail movements and decrease ear movements, a preliminary analysis was performed to see if the type of enrichment could affect the main parameters. This preliminary analysis showed no significant difference in tail movement and ear movement in relation to the type of enrichment material. Nevertheless, it is important to clarify that this result could be due to the low sample size or the observation method. These results do not show that chains are equivalent to other enrichment materials such as straw, as it has already been shown that they are not as adequate as other enrichment materials [29].

One limitation of this study was the sampling time: pigs were observed only 2 min per animal per week, because of practical reasons. Accordingly, we analyzed the data by recording if a pig interacted or did not interact with the enrichment (scored as high or low use) but not the amount of time spent interacting with it.

Additionally, we did not confirm that in that case pig enrichment did produce positive emotions to pigs by using other indicators than tail and ear movements: this could also be a limitation. It may have been better to, firstly, corroborate that animals were experiencing positive emotions when interacting with the enrichment using other indicators, and secondly, measure tail and ear movements to obtain more information about these potential indicators. However, this study opens some perspectives for future investigations that would measure thoroughly these potential indicators and the effects of environmental enrichment on emotions.

Another limitation of this study is that the pigs were sometimes lying down or sitting during the observation period and while using the enrichment materials. In these cases, tail movement could not always be measured because the tail could not be observed or it could not move due to direct contact with the floor, both resulting in the behavior being scored as no tail movement. This factor reduced the values for tail movement duration and frequency and could have directly affected the results. This drawback has also been noted by Reimert and colleagues [9]. One possible solution to reduce resting behaviors during observations in future studies could be to have the observer enter the pen before performing the observations to make the pigs stand up, as was performed in the Welfare Quality Protocol [4].

Another topic to thoroughly explore is the fact that all the pigs in the experiment were tail-docked. It would be interesting in the future to compare tail movement in pigs with and without tail-docking, a practice that should be avoided considering the current European legislation of 2008 [37]. Tail-docking may affect tail movements, as well as communication between animals and social interactions [38]. Additionally, the European Union is exerting pressure to completely fulfill the legislation, and a recommendation was published in 2016 [39]. The use of valid indicators such as tail movements could help farmers prevent tail biting.

No significant difference was found for ear movement frequency in relation to whether enrichment materials were used over all 12 weeks. Previous studies have suggested that ear posture changes in pigs are linked to negative emotions [8]. Other studies have consistently suggested that a high ear movement frequency is a direct indicator of negative emotions by showing that the frequency was significantly higher in barren than in enriched conditions [13,14,17]. These latter studies were performed in controlled conditions with mini-pigs, so the aim of the present study was to investigate whether these results could be reproduced in domestic pigs under experimental conditions. For ear movement, we did not obtain the same results as those obtained with mini-pigs. One hypothesis could be that these pigs did not experience negative emotions, or that the observation method (2 min per animal per week) did not allow observation of this behavior or the emission of negative emotions. Another possible explanation for this difference could be the anatomic difference between the ears of mini-pigs (small in proportion to the head) and Landrace × Large White × Pietrain pigs (very large and heavy in proportion of the head). The auricles of pigs are mobile, and they can move to better detect and locate sound [40]. The anatomical structure of auricles as well as their form could vary depending on the breed, which could also affect ear movement [40]. It is possible that large and heavy ears are less mobile than small ones, as suggested by Wei et al. [41]. Another possible explanation is that the rooms where the fattening pigs were housed are noisier than the rooms where mini-pigs were housed, and this could have affected the ear movements. Finally, the two-minute observation period per week may not have been sufficient to obtain significant results when compared to the observation period in previous studies [13,14,17]. In future research, it would be interesting to investigate these possible explanations, increasing the sampling time, as well as to use different breeds of fattening pigs with different types of ears.

5. Conclusions

This study provides new perspectives on the evaluation of emotions in farm animals and investigates the suitability of using a high tail movement frequency and duration as indicators of emotions with a positive outcome in pigs. Further research is needed to investigate these potential indicators thoroughly in commercial conditions and the hypothesis that enrichment material could produce positive emotions in pigs. This study also investigated the use of a high ear movement frequency as a possible indicator of emotions with a negative outcome, which was shown in previous controlled studies. However, the present study, performed under experimental farm conditions, was not able to confirm this association. Future studies will be planned to continue investigations in this field of research, as well as to study it in sows, which spend more time on the farm and thus have a higher possibility of encountering situations that impact their emotions. The present results may be a first and preliminary step for improving animal welfare assessments of pigs in different breeding systems when measured in a feasible sample of individuals, and for providing a better understanding of their emotions, an important component of animal welfare.

Supplementary Materials: The following are available online at http://www.mdpi.com/2076-2615/9/7/449/s1. File S1: statistical model; Video: Tail movement in tail-docked pigs.

Author Contributions: Conceptualization, M.M.-R., E.F., A.C., C.B.-F., A.V. and P.P.; methodology, M.M.-R.; validation, P.P., A.C. and E.F.; formal analysis, M.M.-R. and E.D.; investigation, M.M.-R. and E.F.; data curation, M.M.-R. and E.D.; writing—original draft preparation, M.M.-R; writing—review and editing, P.P., E.F., A.C., C.B.-F. and A.V.; supervision, P.P., E.F., A.C., C.B.-F. and A.V.; funding acquisition, M.M.-R.

Funding: This research was funded by the 'Convention Industrielle de Formation par la Recherche' (CIFRE) fellowship from the 'Association Nationale de la Recherche et de la Technologie' (ANRT), France.

Acknowledgments: We would like to extend our sincere thanks to all the members of IRTA for giving us the opportunity to perform this study in their facilities and for having the opportunity to work together for some months. In particular, we would like to thank all the team of the Animal Welfare Program and all the members of the Center for the Control and Evaluation of Pigs (Centre de Control i Avaluació de Porcí, Monells). Additionally, we would like to thank the interns Alícia Rodríguez and Roger Vidal for helping us perform the observations. We would also like to thank Céline Lafont-Lecuelle and Sana Arroub for helping us perform the statistical analysis. Finally, we would like to thank all the members of IRSEA's Directory Board and IRSEA's Research and Education Board for giving us the opportunity to develop this project. We also acknowledge the official English language editing company AJE (American Journal Experts) for reviewing the language in our manuscript.

Conflicts of Interest: The authors declare that they have no conflicts of interest to disclose.

References

1. Mellor, D. Animal emotions, behaviour and the promotion of positive welfare states. *N. Z. Vet. J.* **2012**, *60*, 1–8. [CrossRef]
2. Boissy, A.; Lee, C. How assessing relationships between emotions and cognition can improve farm animal welfare. *Rev. Sci. Tech. OIE* **2014**, *33*, 103–110. [CrossRef]
3. Boissy, A.; Manteuffel, G.; Jensen, M.B.; Moe, R.O.; Spruijt, B.; Keeling, L.J.; Winckler, C.; Forkman, B.; Dimitrov, I.; Langbein, J.; et al. Assessment of positive emotions in animals to improve their welfare. *Physiol. Behav.* **2007**, *92*, 375–397. [CrossRef] [PubMed]
4. Welfare Quality. *Welfare Quality® Assessment Protocol for Pigs*; Welfare Quality: Cardiff, UK, 2009; Volume 122.
5. Wemelsfelder, F.; Millard, F.; De Rosa, G.; Napolitano, F. Qualitative behaviour assessment. Assessment of animal welfare measures for layers and broilers. *Welf. Qual. Rep.* **2009**, *9*, 113–119.
6. Mendl, M.; Burman, O.H.; Paul, E.S. An integrative and functional framework for the study of animal emotion and mood. *Proc. R. Soc. B Biol. Sci.* **2010**, *277*, 2895–2904. [CrossRef] [PubMed]
7. Berger, J. Arousal increases social transmission of information. *Psychol. Sci.* **2011**, *22*, 891–893. [CrossRef] [PubMed]
8. Reimert, I.; Bolhuis, J.E.; Kemp, B.; Rodenburg, T.B. Indicators of positive and negative emotions and emotional contagion in pigs. *Physiol. Behav.* **2013**, *109*, 42–50. [CrossRef] [PubMed]
9. Reimert, I.; Fong, S.; Rodenburg, T.B.; Bolhuis, J.E. Emotional states and emotional contagion in pigs after exposure to a positive and negative treatment. *Appl. Anim. Behav. Sci.* **2017**, *193*, 37–42. [CrossRef]
10. Kiley-Worthingthon, M. The tail movement of ungulates, canids and felids with particular reference to their causation and function as displays. *Behaviour* **2011**, *56*, 69–114. [CrossRef]
11. Terlouw, E.M.C.; Porcher, J. Repeated handling of pigs during rearing. I. Refusal of contact by the handler and reactivity to familiar and unfamiliar humans. *J. Anim. Sci.* **2005**, *83*, 1653–1663. [CrossRef]

12. Newberry, R.C.; Wood-Gush, D.G.M.; Hall, J.W. Playful behaviour of piglets. *Behav. Process.* **1988**, *17*, 205–216. [CrossRef]
13. Marcet-Rius, M.; Cozzi, A.; Bienboire-Frosini, C.; Teruel, E.; Chabaud, C.; Monneret, P.; Leclercq, J.; Lafont-Lecuelle, C.; Pageat, P. Selection of putative indicators of positive emotions triggered by object and social play in mini-pigs. *Appl. Anim. Behav. Sci.* **2018**, *202*, 13–19. [CrossRef]
14. Marcet-Rius, M.; Pageat, P.; Bienboire-Frosini, C.; Teruel, E.; Monneret, P.; Leclercq, J.; Lafont-Lecuelle, C.; Cozzi, A. Tail and ear movements as possible indicators of emotions in pigs. *Appl. Anim. Behav. Sci.* **2018**, *205*, 14–18. [CrossRef]
15. Mellor, D.; Patterson-Kane, E.; Stafford, K.J. *The Sciences of Animal Welfare*; UFAW Animal Welfare; John Wiley & Sons: Palmerston North, New Zealand, 2009.
16. Horback, K. Nosing around: Play in pigs. *Anim. Behav. Cogn.* **2014**, *1*, 186–196. [CrossRef]
17. Marcet-Rius, M.; Kalonji, G.; Cozzi, A.; Bienboire-Frosini, C.; Monneret, P.; Kowalczyk, I.; Teruel, E.; Codecasa, E.; Pageat, P. Effects of straw provision, as environmental enrichment, on behavioural indicators of welfare and emotions in pigs reared in an experimental system. *Livest. Sci.* **2019**, *221*, 89–94. [CrossRef]
18. EFSA AHAW Panel (EFSA Panel on Animal Health and Welfare). Scientific Opinion concerning a multifactorial approach on the use of animal and non-animal-based measures to assess the welfare of pigs. *EFSA J.* **2014**, *12*, 1–101.
19. Van de Weerd, H.A.; Day, J.E. A review of environmental enrichment for pigs housed in intensive housing systems. *Appl. Anim. Behav. Sci.* **2009**, *116*, 1–20. [CrossRef]
20. Boissy, A.; Aubert, A.; Désiré, L.; Greiveldinger, L.; Delval, E.; Veissier, I. Cognitive sciences to relate ear postures to emotions in sheep. *Anim. Welf.* **2011**, *20*, 47.
21. Reefmann, N.; Kaszàs, F.B.; Wechsler, B.; Gygax, L. Ear and tail postures as indicators of emotional valence in sheep. *Appl. Anim. Behav. Sci.* **2009**, *118*, 199–207. [CrossRef]
22. Reefmann, N.; Wechsler, B.; Gygax, L. Behavioural and physiological assessment of positive and negative emotion in sheep. *Anim. Behav.* **2009**, *78*, 651–659. [CrossRef]
23. Freymond, S.B.; Briefer, E.; Zollinger, A.; Gindrat-von Allmen, Y.; Wyss, C.; Bachmann, I. Behaviour of horses in a judgment bias test associated with positive or negative reinforcement. *Appl. Anim. Behav. Sci.* **2014**, *158*, 34–45. [CrossRef]
24. Fraser, D.; Philips, P.A.; Thompson, B.K.; Tennessen, T. Effect of straw on the behaviour of growing pigs. *Appl. Anim. Behav. Sci.* **1991**, *30*, 307–318. [CrossRef]
25. Van de Weerd, H.A.; Baumans, V. Environmental enrichment in rodents. Environmental enrichment information resources for laboratory animals. *AWIC Resour. Ser.* **1995**, *2*, 145–149.
26. O'Connell, N.E.; Beattie, V.E. Influence of environmental enrichment on aggressive behaviour and dominance relationships in growing pigs. *Anim. Welf.* **1999**, *8*, 269–279.
27. Douglas, C.; Bateson, M.; Walsh, C.; Bédué, A.; Edwards, S.A. Environmental enrichment induces optimistic cognitive biases in pigs. *Appl. Anim. Behav. Sci.* **2012**, *139*, 65–73. [CrossRef]
28. Council Directive 2001/88/EC of 22 October 1901 amending Directive 91/630. ECC Laying Down Minimum Standards for the Protection of Pigs. *Off. J. Eur. Communities* **2001**, *21*, 316–320.
29. European Union. *Commission Recommendation (EU) 2016/336 of 8 March on the Application of Council Directive 2008/120/EC Laying Down Minimum Standards for the Protection of Pigs as Regards Measures to Reduce the Need for Tail-Docking*; Official Journal of the European Union: Brussels, Belgium, March 2016.
30. Studnitz, M.; Jensen, M.B.; Pedersen, L.J. Why do pigs root and in what will they root? A review on the exploratory behaviour of pigs in relation to environmental enrichment. *Appl. Anim. Behav. Sci.* **2007**, *107*, 183–197. [CrossRef]
31. Fàbrega, E.; Marcet-Rius, M.; Vidal, R.; Escribano, D.; Cerón, J.J.; Manteca, X.; Velarde, A. The Effects of Environmental Enrichment on the Physiology, Behaviour, Productivity and Meat Quality of Pigs Raised in a Hot Climate. *Animals* **2019**, *9*, 235. [CrossRef]
32. Martin, P.; Bateson, P. *Measuring Behaviour, an Introductory Guide*, 3rd ed.; Cambridge University Press: Cambridge, MA, USA, 2007.
33. Reimert, I.; Bolhuis, J.E.; Kemp, B.; Rodenburg, T.B. Emotions on the loose: Emotional contagion and the role of oxytocin in pigs. *Anim. Cogn.* **2015**, *18*, 517–532. [CrossRef]

34. Salaun, M.C.; Val-Laillet, D. Study of animal behavior and welfare of miniature pigs in France. In *Proceedings Bilateral Symposium on Miniature Pigs for Biomedical Research in Taiwan and France, 2013*; Livestock Research Institute, Council of Agriculture: Taiwan, China, 2013.
35. Ellegaard, L.; Cunningham, A.; Edwards, S.; Grand, N.; Nevalainen, T.; Prescott, M.; Schuurman, T. Welfare of the minipig with special reference to use in regulatory toxicology studies. *J. Pharmacol. Toxicol Methods* **2010**, *62*, 167–183. [CrossRef]
36. Tuyttens, F.A.M. The importance of straw for pig and cattle welfare: A review. *Appl. Anim. Behav. Sci.* **2005**, *92*, 261–282. [CrossRef]
37. Council Directive 2008/120/EC of 18 December 2008 laying down minimum standards of the protection of pigs. *Off. J. Eur. Union* **2008**, *7*, 5–13.
38. Houpt, K.A. *Domestic Animal Behaviour for Veterinarians and Animal Scientists*, 5th ed.; Wiley-Blackwell Pub.: Ames, IA, USA, 2005.
39. European Commission. *Cutting the Need for Tail-Docking 2016*; European Commission: Brussels, Belgium, 2016.
40. König, H.E.; Liebich, H.G. *Anatomia dos Animais Domésticos: Texto e Atlas Colorido*; Artmed®Editora S.A.: Sao Paolo, Brazil, 2004; Volume 2, pp. 309–323.
41. Wei, W.H.; De Koning, D.J.; Penman, J.C.; Finlayson, H.A.; Archibald, A.L.; Haley, C.S. QTL modulating ear size and erectness in pigs. *Anim. Genet.* **2007**, *38*, 222–226. [CrossRef] [PubMed]

© 2019 by the authors. Licensee MDPI, Basel, Switzerland. This article is an open access article distributed under the terms and conditions of the Creative Commons Attribution (CC BY) license (http://creativecommons.org/licenses/by/4.0/).

Article

The Effects of Environmental Enrichment on the Physiology, Behaviour, Productivity and Meat Quality of Pigs Raised in a Hot Climate

Emma Fàbrega [1,*], Míriam Marcet-Rius [2], Roger Vidal [1], Damián Escribano [3,4], José Joaquín Cerón [3], Xavier Manteca [4] and Antonio Velarde [1]

[1] Animal Welfare Program, IRTA, Veïnat de Sies s/n, 17121 Monells, Spain; cvr.milan42@gmail.com (R.V.); antonio.velarde@irta.cat (A.V.)
[2] Physiological and Behavioural Mechanisms of Adaptation Department, IRSEA (Research Institute in Semiochemistry and Applied Ethology), Quartier Salignan, 84400 Apt, France; m.marcet@group-irsea.com
[3] Interdisciplinary Laboratory of Clinical Analysis (Interlab-UMU), Regional Campus of International Excellence 'Campus Mare Nostrum', University of Murcia, Campus de Espinardo s/n, Espinardo, 30100 Murcia, Spain; det20165@um.es (D.E.); jjceron@um.es (J.J.C.)
[4] Department of Animal and Food Science, School of Veterinary Science, Autonomous University of Barcelona, 08193 Bellaterra (Barcelona), Spain; xavier.manteca@uab.cat
* Correspondence: emma.fabrega@irta.cat

Received: 26 March 2019; Accepted: 9 May 2019; Published: 13 May 2019

Simple Summary: European Union (EU) legislation states that the routine tail docking of pigs should not be carried out and that manipulable materials should be made available to all pigs to prevent tail biting and allow them to behave naturally. However, between 90 and 95% of pigs within the EU still have their tails docked to avoid the risk of tail biting. Farmers say they require information tailored to their particular production systems before they abandon this practice. In this study, four types of enrichment materials used in Spanish pig fattening production systems are compared. Most of these systems have fully slatted floors and high external temperatures for considerable periods of the year. The effects of chains (the control group), wood, paper or straw in a rack on the behaviour, health/physiology, performance and meat and carcass quality are evaluated. Straw in a rack was found to be the best material to meet the behavioural needs of pigs, whereas paper met the criteria of being manipulable, but only for a short period. To avoid the risk of blockages in the slurry system, there are some practical issues to consider and improvements to be made to the design of the rack for providing straw used in this study.

Abstract: Some positive effects regarding the use of enrichment material on the stimulation of pig exploration and a reduction in redirected behaviour was reported. This study aims to evaluate the effects of four enrichment materials on the behaviour, physiology/health, performance and carcass and meat quality in pigs kept in Spanish production conditions. Ninety-six male pigs (six pigs/pen) ranging from 70 to 170 days old were used. Chains were used for the control group (CH), and wooden logs (W), straw in a rack (S) or paper (P) were also used. The pigs were subjected to two pre-slaughter treatments: 0 or 12 h of fasting. Their behaviour was observed for 12 weeks using scan and focal sampling. Samples of the Neutrophil: Lymphocyte (N:L) ratio and lactate were obtained from the pigs at 66 and 170 days old. Saliva samples for Chromogranin-A (CgA) were obtained at 67, 128, 164 and 170 days old. The weight, skin lesions and feed intake of the pigs were recorded. S triggered more exploratory behaviour than W and CH ($P < 0.001$). Skin lesions and redirected behaviour were lower for pigs with S ($P < 0.01$ and $P < 0.05$, respectively). The pigs offered S presented lower CgA after no fasting than pigs with P or CH ($P = 0.055$). Lactate was higher in pigs with W and CH treatments, regardless of fasting ($P < 0.05$). The N:L ratio increased over time ($P < 0.05$). No other significant effects were found. Overall, straw in a rack was the enrichment material that enhanced pig inherent behaviour.

Keywords: environmental enrichment; pig; behaviour; performance; Chromogranin-A; lactate; skin lesions; meat quality

1. Introduction

Environmental enrichment has been defined as the improvement in the biological functioning of captive animals resulting from modifications to their environment [1]. Enriched environments aim to improve the welfare of animals by allowing them to perform more of their species-specific behaviour. Pigs kept in intensive production systems are exposed to several management or environmental conditions which do not always meet their individual requirements. Specifically, pigs in many intensive husbandry systems are often not provided with the proper foraging materials which can lead to stress, because they normally spend most of their active time rooting if given the possibility [2]. The lack of proper enrichment materials to allow them to express their foraging behaviour may also lead to frustration (for a review see [3–5]) and has been found to be an underlying cause of tail biting [6–9]. The practice of tail docking has been adopted for a long time and is common in pig farms to prevent the risk of tail biting, although it does not always work and outbreaks may still occur in docked pigs [4,8,9].

The current European Union (EU) Directive for the protection of pigs [10] stipulates that enrichment materials, such as straw or other suitable materials, should be provided to satisfy the behavioural needs of all categories of pigs (i.e., to satisfy both rooting and nesting behavioural needs). Furthermore, the directive [10] specifies that tail docking should not be routinely carried out and stipulates the measures that farmers should take before they resort to docking. Although the legal and social demands to avoid tail docking are increasing, recent studies show that 90–95% of pigs in the EU still have their tails docked to avoid the risk and consequences of tail biting [11,12]. Farmers say they are reluctant to stop tail docking for the following reasons: (1) the solutions are not straightforward because the reasons why an outbreak of tail biting develops, even in docked pigs, are complex (2) the fear and frustration of failure if they move to raising pigs with intact tails; (3) the economic consequences of both the outbreaks and of implementing prevention measures that may not work. Therefore, the EU is investing its efforts in disseminating the existing information on how to prevent tail biting and conducting research into new strategies, such as innovations in environmental enrichment. One important consideration is how the existing knowledge can be applied in various farming and climatic conditions in different countries. One of the objectives of this study is to evaluate the effects of different enrichment materials on docked pigs in the Spanish climate, as a first step towards increasing farmers' confidence in strategies that could allow them to move towards raising pigs with intact tails.

It has been suggested that the availability of enrichment materials, especially straw, may increase the general activity, reduce stereotypes and increase the social interactions of the pigs [3,13]. Moreover, other positive effects have been reported, such as a reduced incidence of belly nosing, tail and ear biting, improved learning capabilities and reduced fearfulness [13–16]. Van de Weerd and Day [3] suggest that to be successful the enrichment material should meet four criteria: (1) it should increase species-specific behaviour, (2) it should maintain or improve levels of health, (3) it should improve the economics of the production system, and (4) it should be practical to employ. Scientific literature suggests that straw and similar substrate materials provided on the floor have the greatest potential to meet the first criteria, which addresses pig welfare [3,17]. However, in Spain most fattening pigs are kept on fully or partially slatted floors, and the liquid manure systems limit the use of substrate enrichments due to the possibility of blockages in the slurry system (i.e., the practicality criteria would not be met). Moreover, the high temperature conditions may prevent the use of large amounts of straw during certain periods of the year as this could cause heat stress. Therefore, point source object enrichment (i.e., 'objects limited in size and restricted to certain pen location(s)', as defined in [3]) may be more appropriate for slatted or partially slatted flooring production systems. Some studies have presented the potential beneficial effects of using point source materials [13,18,19], but their effectiveness varies depending on the type and properties of the material provided, as well as its position, the sense of novelty, cleanliness, interaction with management or climatic conditions and other attributes [3]. Some studies in fattening pigs have demonstrated that the use of ample, fresh wood attached horizontally to chains, as used in this study, can increase object manipulation and reduce both tail and ear damage when compared with single chains, branched chains or plastic objects [18]. Research into the effects of enrichment materials has often focused on behavioural outcomes, but it is also important to evaluate other aspects such as their effect on physiology, health, productivity, economic viability and practicability for farmers, and even the potential influence on the product quality [3,20]. Most objects have been considered insufficient on their own in having a positive effect on all the parameters. The EU Recommendation 2016 [21] classifies enrichment material as "optimal", "suboptimal" or "marginal", according to the level of interest, and provides guidance on how to combine them to optimise their efficacy. However, the list is not exhaustive and many aspects of how to enrich pens properly, especially fully slatted ones, are yet to be revealed.

The aim of this study is to evaluate the effect of four different types of enrichment material, feasible to implement in Spanish pig housing and climate conditions on: (1) behaviour; (2) physiology and health indicators; (3) productivity and (4) carcass and meat quality, in fattening docked pigs. The four criteria suggested by Van der Weerd and Day [3] have been used to evaluate the possibility of success of the four different types of enrichment material in the Spanish pig husbandry scenario.

2. Materials and Methods

2.1. Ethical Considerations

The housing, husbandry and use of animals for the procedures described in this paper was carried out in accordance with Spanish (RD 53/2013) and European legislation regarding animal experimentation and the care of animals under study. The project, including the experimental procedure, was approved by the Institute of Agri-food Research and Technology (IRTA) Ethics Committee and received a grant and approval (n° AGL2015-68373-C2-2-R) from the Spanish Ministry of Economy and Enterprises (MINECO). The authors acknowledge that the Commission staff working document accompanying the EU Recommendation 2016 [21] classifies the materials used in this experiment as suboptimal or marginal. This means that according to the commission working document, chains are considered as marginal enrichment materials, and therefore to fulfil the behavioural needs of the pigs on the farm must be complemented with either optimal or suboptimal materials. In this experiment, permission from the ethical committee was obtained to use chains in the 'control' group. Other materials used in this experiment come under the category of suboptimal to marginal, as 'wood' is considered suboptimal or marginal depending on its softness, as are paper and straw in a rack. According to the EU Recommendation 2016 [21], all these materials should be combined with other materials to complement their properties. However, the aim of this study is to evaluate the effects of these materials when isolated, and to avoid possible confounding results from their combination. A preliminary trial was conducted as part of this research which indicated that using some of the enrichment items individually under similar conditions, led to a considerable occurrence of tail biting. Therefore, due to ethical concerns, the authors decided to compare the efficacy of those materials (wood, chains) with new ones (straw in a rack) in docked pigs, as the first step towards moving to undocked pigs.

2.2. Animals and Housing and Experimental Conditions

Ninety-six male pigs ((Landrace, Large white) and Pietrain) were used in this experiment. The piglets were all born in a commercial pig farm within five days, from which two piglets/litter of similar weight were chosen. All pigs had their tails docked before they were seven days old. The pigs selected for this study arrived at the experimental farm with an average live weight of 23.50 ± 2.5 kg (mean ± SD) and were 60 days old. They were kept together before the start of the experiment until they were 70 days old and had an average weight of 26.1 ± 3.00 kg. The pigs were randomly divided into four treatments (24 pigs/treatment), as explained in Section 2.3. They were housed in groups of six in two identical rooms, with eight pens in each room (i.e., two pens of each enrichment type in each room). For differentiation purposes, each pig was allocated a different colour tag (blue, yellow, red, orange, green and white). The pigs were also identified by means of an electronic ear transponder. The pens were identical except for the enrichment material provided. The stocking density was 0.9 m^2/pig, and the floors were fully slatted (commercial size: 80 mm wide slats with an 18 mm gap between them). There was one bowl-type drinker and one hopper available. The pens were divided by metallic fences which allowed the pigs to see each other and there was a solid metallic barrier in the lower part of the fence to prevent the enrichment materials moving between pens. The room was controlled by an automatic control system that regulated both temperature and ventilation. The room temperature was maintained at 22 ± 5 °C. The pigs were fed on a commercial pig diet provided ad libitum and they had continuous access to drinking water. This diet consisted of a commercial concentrate provided in a phased feeding regime (15.04% crude protein and 2.321 Kcal net energy at mid fattening). The experiment lasted for approximately three months, starting at the end of February and ending in mid-June, when the outside maximum temperature had already reached 30 °C.

2.3. Experimental Treatments

The four enrichment materials provided were chosen because they were feasible for use with fully slatted floors and in hot climates, and on the assumption that their practicality would encourage farmers to use them and combined when necessary. They consisted of (a) chains (considered as the control group); (b) wooden logs attached to a chain; (c) straw; and, (d) paper. For the chain treatment, two chains (50 cm long) were attached perpendicularly to the metal fence bars of each pen. One was attached to the door and another was attached to the same fence at a distance of approximately 50 cm. For the wood treatment, two wooden logs (30 × 5 × 5 cm) were hung perpendicularly from a chain, in the same positions as described for the chain treatment. Both the chains and wooden logs were near the floor but did not touch it. The wooden logs were made from fresh wood from the ash tree and were replaced after 20 days of use. Non-chopped straw was continuously available in a rack (60 × 40 × 80 cm), without a tray underneath to collect any spare straw. New straw was added once a day in the morning to ensure that the racks were completely full. For the paper treatment, two entire newspaper sheets (40 × 30 cm) were provided for each pig every day, one in the morning at around 8:30h and the other at midday around 14:00h.

2.4. Behavioural Observations

Instantaneous scan sampling and continuous focal sampling as described by Martin and Bateson [22] was used to record behaviour using an ethogram (as illustrated in Table 1). Once a week, for 12 weeks, two trained observers carried out direct observations of all the pens in each room in sessions of approximately four consecutive hours (from 10:00h. to 14:00h). Behavioural observations started when the pigs were 77 days old (one week after the provision of the enrichment treatments) and ended at 164 days (i.e., after three months of observations). Before starting the experimental procedure, the observers carried out preliminary trials to test their repeatability (a coefficient of correlation r > 0.8, $P < 0.001$ at the start of the procedure). The two observers carried out their observations simultaneously, one in each room, to avoid possible differences of time between rooms. Each group of four pens was scanned at seven-minute intervals, and focal samplings were recorded at three minutes per pen between two consecutive scan samplings (i.e., instantaneous scan samplings of activity were performed for two minutes at the beginning and end of each observation round and three minutes of focal sampling was performed in between, with two minutes rest). The same procedure was carried out for the other group of four pens in the same room. This scan and focal sampling interval was adapted from a similar methodology used in a previous study to suit the conditions of this experiment [13]. Thus, each observation day provided a total of 12 scans per animal and nine minutes of focal sampling per pen, divided into three periods of three minutes. Every week, the observers swapped rooms with each other and changed the order of observation of the eight pens within each room so they each observed all the rooms at different times. The behaviour observed according to each observational methodology is summarised in Table 1. Ten minutes before starting the observations, each observer entered the room and walked around for five minutes to allow the pigs to get used to their presence. Then, the observer moved towards the centre of the room, and remained there for another five minutes before starting the observation. The observation was carried out from the centre of the room from where all the pens could be seen.

Table 1. Ethogram of the Behaviours Evaluated in the Scans or Focal Samplings taken over 12 weeks of Observations.

Specific Category	Definition
SCAN SAMPLING	
Activity	
Standing inactive	Pig is upright on all four legs, neither moving forward or backward
Walking	Pig is upright on all four legs, and moves in the pen
Sitting inactive	Pig is upright on two front legs, and hindquarter
Lying	Pig is recumbent on its belly or side
Behaviour	
Eat/Drink	Head or snout over bowl or trough
Interaction with the pen	Licking, chewing, nosing or sniffing unanimated objects from the pen, excluding enriched material
Interaction with enrichment material	Interaction with each type of enrichment material
Social interaction	Head or snout in contact with another pig (includes any positive or negative social behaviour)
Other behaviour	Behaviour different to that previously described
FOCAL SAMPLING	
Positive social behaviour	Head or snout in contact with another pig. The receiver doesn't react negatively
Negative social behaviour	Aggression, fights to another pig with a negative response from the receiver
Stereotypies	Stereotypical behaviour, basically animal rubs the bars of the crate with the mouth, and other stereotypical behaviour
Redirected behaviour	Vertical rub movements with the snout to the belly of a pen-mate (belly nosing), bite or suck the ear of a pen-mate (ear biting) and bite or suck the tail of a pen-mate (tail biting)
Interaction with enrichment material	Interaction with each type of enrichment material
Sexual behaviour	Sexual behaviour, one animal mounting or trying to mount another pig
Other behaviour	Any other behaviour

2.5. Physiological Parameters

Saliva samples were obtained to analyse salivary Chromogranin-A (CgA), which has previously been reported as an indicator of acute and/or chronic stress [23,24]. The first saliva sample was taken when the pigs were 67 days old, three days before the enrichment materials were provided, to measure the basal levels. Two other samples were obtained at 128 and 164 days of age (approximately mid-procedure and end, and 1.5 and three months after providing the enrichment materials). A fourth saliva sample was collected at the slaughterhouse (170 days old), during lairage. Saliva samples were taken between 08:00 h and 11:00 h., by allowing the animals to chew on a cotton bud (Salivette®; SARSTEDT AG & Co., Nürbrecht, Germany) for more than 30 seconds and randomising the animals from different treatments. Cotton buds were offered using a clamp, with no restraint of the pigs. The sampling time was fixed up to a maximum of five minutes. Once sampled, and before the analysis, the salivettes® were centrifuged at 3500 rpm for 10 min, the cotton buds were removed, and the tubes were frozen at −20 °C. Salivary CgA was measured using a validated time-resolved immunofluorometric assay (TR-IFMA) [23,24].

Blood samples were collected to perform a hemogram and determine the ratio neutrophil: lymphocytes (N:L) as an indicator of the immune system response and potential indicator of stress. Lactate was also measured in the second blood sample. The basal sample of blood was collected at 66 days of age (four days before providing enrichment) via a jugular vein puncture between 08:00 and 11:00 a.m. The pigs were gently restrained by one trained operator to obtain the samples, and the procedure took less than five minutes for all pigs. Samples were obtained at 170 days of age just after exsanguination at the slaughterhouse. The blood samples were collected in ethylene diamine tetra acetic acid (EDTA) (1 mg/mL) coated test tubes. Once sampled, the blood was refrigerated and immediately sent to the laboratory for basic haematology analysis. Another blood sample was collected at the slaughterhouse in EDTA tubes and immediately used to measure lactate using the commercial kit Lactate Scout-4® (Biolaster, Gipuzkoa, Spain).

2.6. Performance Indicators and Body Lesions

The weight of the pigs was recorded when they were 68 days old (before starting the experiment), and then at 90, 107, 129, 157 and 168 days using a cage with a scale (MBWA100 Meier-Brakenberg; GmbH & Co, Extertal, Germany). Feed intake was recorded at pen level using a trolley equipped with scales (VLIEBO, Veenendal, The Netherlands), which allowed the amount of food provided to each pen to be recorded. When the pigs were weighed individually, the food left in the hopper was also weighed to allow for estimates of food intake by periods. The feed conversion ratio was estimated from the mean average daily weight gain and mean daily feed intake per pen (i.e., total pen feed intake/6 pigs per pen).

Skin lesions were assessed the day before the pigs were weighed in their home pens. The total number of lesions on one side of the pig in each one of five regions (ears, front, middle, rear -including tail- and legs) as defined in the Welfare Quality® [25] was recorded.

2.7. Fasting Treatments and Carcass and Meat Quality Parameters

All the animals were slaughtered in a commercial abattoir at the age of 24 weeks (170 days). The mean live weight was 114.92 ± 11.02 kg. All pigs were gas stunned before sticking. The pigs were subjected to two different periods of on farm fasting, to provide different pre-slaughter stress conditions. The pigs in half of the pens per treatment, equally distributed in the two rooms, were fasted for 12 h and half were not fasted (i.e., two pens/enrichment and fasting treatment, 12 pigs/enrichment and fasting treatment). Transport to the slaughterhouse took 30 min and the pigs were kept in lairage for two hours before slaughter. They were gently handled when being loaded and unloaded to avoid mixing with unfamiliar pigs in the truck. The carcass weight was measured at 45 min *post mortem*, and the carcass yield was calculated by dividing the carcass weight by the live body weight. Furthermore, the backfat (LR3/4FOM) and muscle thicknesses (MFOM) were measured between the third and fourth last ribs, at 6 cm from the midline, using a Fat-O-Meat'er probe (Frontmatec A/S, Odense, Denmark). The official Spanish equation was used to calculate the lean percentage from the backfat and muscle thickness measurements. A ruler was used 24 h *post mortem* to measure the left carcass of each animal: (a) minimum fat and (b) skin thickness (perpendicular to the skin) over the *gluteus medius* muscle (MLOIN). A tape was used to measure the loin (from the atlas to the first lumbar vertebra) and carcass length (the first rib to the anterior edge of the pubic symphysis).

A Crison portable meter (Crison, Barcelona, Spain) equipped with a Xerolyte electrode and a Pork Quality Meater (PQM-I, INTEK Aichach, Germany) were used 24 h *post mortem* to measure muscle pH and electrical conductivity in the muscle *Longissimus thoracis* (LT) at the last rib level and Semimembranosus (SM), respectively. Luminosity L*, tendency to red a* and tendency to yellow b* (colour parameters on the CIELab space) were obtained using a Minolta Chromometer (CR-400, Minolta Inc., Osaka, Japan).

2.8. Statistical Analysis

All the statistical analyses were conducted using the Statistical Analysis System (SAS version 9.2; SAS institute Inc., Cary, NC; USA). Significance was established at $P < 0.05$, and the exact P value is provided for all those results with a $P < 0.1$. The results are presented as mean ± standard deviation (SD) or SE (standard error) unless otherwise indicated. The Shapiro-wilk test (with PROC UNIVARIATE in SAS) was used to examine the normality of the distributions. The experimental unit for the lesions, physiological indicators and carcass and meat quality parameters was individual, whereas for the behavioural observations and productivity data (weight, feed intake and food conversion ratio) the pen was considered as the experimental unit. The least square means of fixed effects (LS-means) with Tukey adjustment was used for comparisons when analysis of the variance indicated differences ($P < 0.05$).

The productivity parameters and physiological indicators presented a parametric distribution and were analysed using a generalised linear mixed model for repeated measurements (MIXED procedure and the covariance matrix Autoregressive AR(1) in SAS). The AR(1) covariance matrix was chosen because according to Schwarz's Bayesian criterion it was the best fit for the model. The enrichment treatment and time and the interaction were considered as fixed effects, while the pen was considered as a random effect for the physiological indicators. For the weight analysis, the weight before starting the experiment was considered a covariate. The carcass and meat quality parameters were presented as a parametric distribution and were analysed using a general linear model (GLM), using enrichment treatment, fasting time and their interaction as fixed factors. Carcass weight was considered a covariate in those parameters where it was proved to have a significant effect. The lactate and CgA data collected at the slaughterhouse after two on farm fasting periods was also analysed using a GLM model with enrichment type, fasting time and its interaction as fixed effects.

The skin lesions and behavioural data were analysed using non-parametric generalised linear models for repeated measurements (GLIMMIX or GENMOD procedure, respectively). The Poisson distribution was used and for some parameters a negative binomial, depending on the deviance [26]. Lesions before providing enrichment were considered as a covariate. Enrichment material and time, and its interaction were used as fixed effects, and the pen as a random effect for skin lesions. The behavioural data from the scan samples was analysed as a percentage of the scans in each category in relation to the total number of scans per day, while data from focal sampling was analysed as the total number of counts observed in each category per pen and day. In the first analysis all 12 observations over time were included as a time effect on the behavioural data using a non-parametric generalised linear model (GENMOD procedure) with repeated measurements and the covariance matrix AR(1). An effect of time was observed, but the effect was the same as when averaging the data in three periods (i.e., four observations per period). Therefore, for simplicity, the results are presented for the averaged results in three periods, and not for the 12 observations over time.

3. Results

3.1. Behavioural Observations

Considering the four treatments, time had a significant effect on the lying behaviour, being a percentage higher in the third period compared to the other two periods in all four treatments (average % for all treatments: Period 1 = 46.90 ± 12.25; Period 2 = 47.51 ± 15.25 and Period 3 = 52.34 ± 15.70, χ^2 (Chi-Square) (6, N = 96) = 2.45, $P = 0.0049$). Moreover, more differences were observed when considering the time*treatment interaction. The interaction with the enrichment material significantly increased in the second period for pigs in the straw treatment compared to the first period. Whereas for pigs in the chain treatment the percentage of interaction decreased in the second and third periods compared with the first (Figure 1a, χ^2 (4, N = 72) = 8.84, $P = 0.009$). Activity was significantly higher for pigs in the paper treatment in the first and final periods, whereas a higher activity percentage was presented in the first period compared to the other three treatments (Figure 1b, $\chi2^2$(6, N = 96) = 8.65, $P = 0.009$). The pigs in the wood treatment showed a decrease in activity over time (χ^2(6, N = 96) = 8.79, $P = 0.014$). Social behaviour was higher in the first period for the pigs in the straw treatment compared with the other three treatments, whereas in the final period a higher percentage was found for pigs in the paper treatment (Figure 1c, χ^2 (6, N = 96) = 7.10, $P = 0.06$). Interaction with the pen structures was higher for the pigs in the paper treatment compared with the other three treatments in all three periods (Figure 1d, χ^2 (6, N = 96) = 7.82, P = 0.07). The eating/drinking percentage increased for the pigs in the paper and straw treatments in the third period compared with that in the first and second periods, but this tendency was not found for pigs in the chain and wood treatments (12.83 ± 6.06, 12.84 ± 6.77, 10.15 ± 5.18 and 10.50 ± 5.04, for the paper, straw, chain and wood treatments in P3, respectively, P = 0.06).

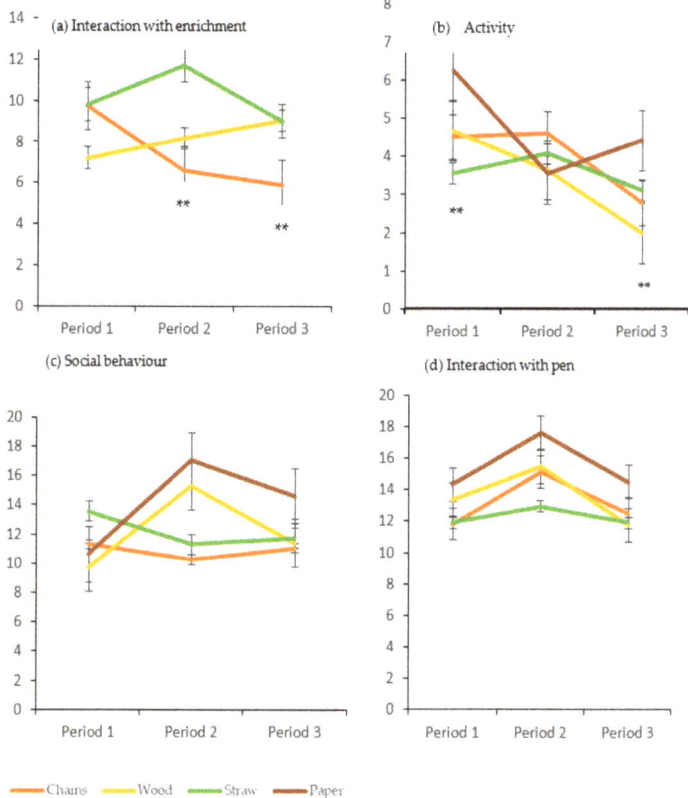

Figure 1. Behaviours observed by scan sampling (Mean percentage and SE) for the 4 types of enrichment materials. 1-3: averaged observations for first, second and third month of observations. ** P < 0.01 (see text for explanation of statistical differences).

For the focal observations, the treatment*time interaction was significant for the use/interactions with the enrichment materials. This showed a decrease over the three periods of observation for pigs in the chain treatment whereas for pigs in the straw treatment, this interaction was significantly higher in the second period (F (Fisher) (4, 18) = 9.22, P = 0.0003, Figure 2a). For redirected behaviour, the counts were lower in the first period in the straw treatment and then they decreased, particularly between the second and third period. Whereas for the pigs in the chain and wood treatments, the decrease was only pronounced between the first and second periods (Figure 2b, F (6, 24) = 7.01, P = 0.025). There was an increase over time in negative social behaviour for pigs in the paper treatment (12.50 ± 5.00, 9.75 ± 2.7, 8.00 ± 4.76 and 8.50 ± 2.38, for the paper, chain, wood, and straw treatments at P3, respectively, P = 0.07). Sexual behaviour and positive social behaviour were not significantly affected by either enrichment treatment, time or its interaction.

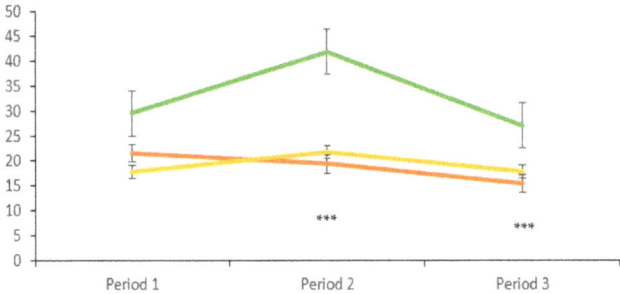

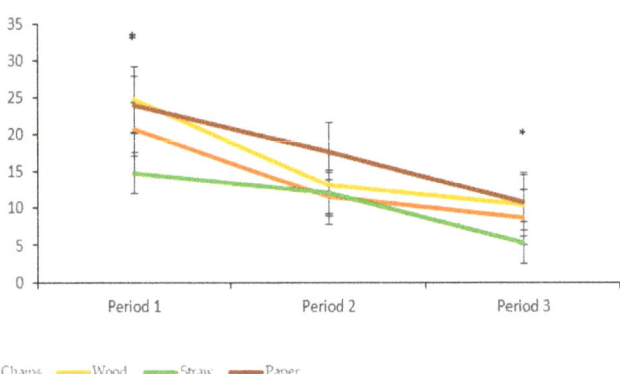

Figure 2. The behaviour observed by focal sampling (Mean counts and SE) for the four types of enrichment materials. 1-3: averaged observations for first, second and third month of observations. * $P < 0.05$, *** $P < 0.001$ (see text for explanation of statistical differences).

Due to the low incidence, the differences in stereotypies could not be statistically analysed. However, the stereotypies were counted on 38 occasions during the focal observations, 63.5% (24 counts) in the paper treatment, 28.9% (11 counts) in the chain treatment, 5.2% in the wood treatment (2 counts) and 2.6% (one count) in the straw treatment.

3.2. Physiological Parameters

The CgA level was measured when the pigs were 128 days old (the first sample was taken after provision of enrichment) and was significantly higher compared to the basal level at 67 days old and the level in the last sample at 164 days old (F (2, 259) = 0.2, $P = 0.006$ Table 2). Pigs in the paper treatment showed a higher ratio of increase of CgA at 128 days old compared with that for the other three treatments (5.06 ± 4.80 vs. 2.59 ± 2.15, 1.24 ± 0.53 and 2.0 ± 2.04, for the paper, chain, wood and straw treatments, respectively, F (1, 164) = 5.31, $P = 0.07$). The N:L ratio was significantly higher in the second sample compared to the basal level before the provision of enrichment (F (1, 171) = 7.33, $P = 0.008$). There was no significant effect of fasting or interaction fasting*enrichment treatment found for the ratio N:L.

Table 2. The Mean (and standard deviations) of the various physiological indicators of welfare for the 4 types of enrichment over time.

	Enrichment Material Treatment				P Treatment [3]	P Time
	Chain	Wood	Straw	Paper		
CgA 67 [1]	0.77 [a] ± 0.87	0.73 [a] ± 0.99	1.10 [ab] ± 1.57	1.08 [a] ± 1.57		
CgA 128	1.15 [b] ± 2.20	1.09 [ab] ± 1.64	1.34 [b] ± 2.19	2.51 [b] ± 3.89	NS	**
CgA 164	0.73 [a] ± 0.61	0.72 [a] ± 0.52	0.63 [a] ± 0.57	0.71 [a] ± 0.44		
N:L 66 [2]	0.66 [a] ± 0.32	0.67 [a] ± 0.24	0.67 [a] ± 0.32	0.62 [a] ± 0.21	NS	**
N:L 170	0.79 [b] ± 0.31	0.83 [b] ± 0.31	0.82 [b] ± 0.45	0.77 [b] ± 0.47		

[1] CgA 67, 128, 164 = Chromogranin A (μg/ml) at 67, 128 or 164 days old. [2] N:L 66, 170 = Neutrophil/Lymphocyte ratio at 66 or 170 days old. [3] [a], [b] and [c]: means with different superscripts present significant differences due to treatment (row comparisons) or due to time (column comparisons). ** = $P < 0.01$.

The pigs provided with the straw enrichment showed lower levels of CgA when subjected to 0 h of on farm fasting compared to the pigs in the other three enrichments ($F (3, 85) = 1.74$, $P = 0.055$, Figure 3). After 12 h of on farm fasting, the pigs in the wood treatment presented with lower levels of CgA, compared to the pigs in the paper treatment. Pigs in the chain and straw treatments presented with values between ($F (3, 85) = 1.74$, $P = 0.055$).

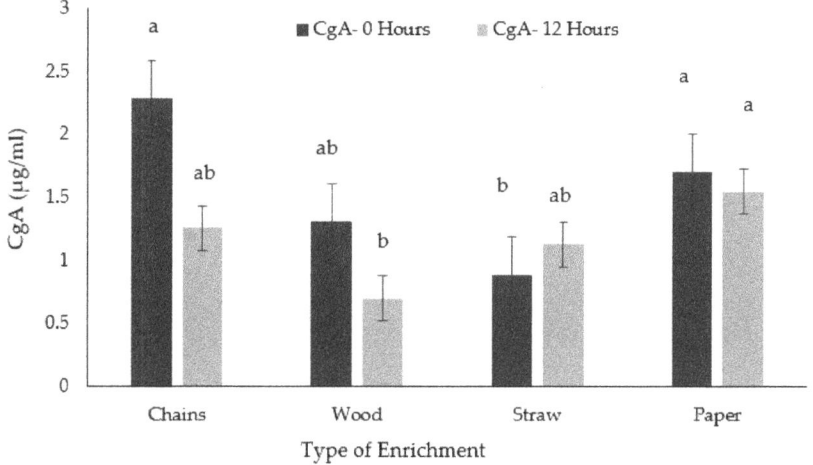

Figure 3. Means (and standard errors) of Chromogranin A for the 4 types of enrichment materials after the two different fasting times (0 and 12 h). Different superscripts indicate significant differences due to treatment*fasting interaction.

The pigs provided with straw and paper presented significant lower levels of lactate compared to pigs in the chain and wood treatment, both after 0 and 12 h of on farm fasting ($F (7,87) = 3.49$, $P = 0.019$, Figure 4).

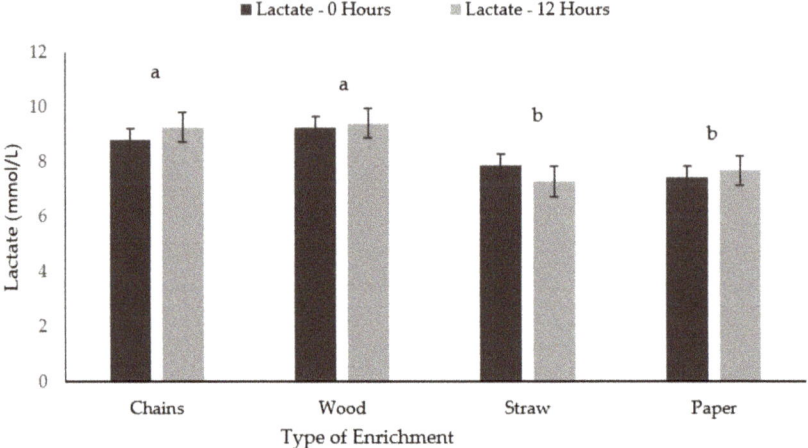

Figure 4. Means (and standard errors) of Lactate for the four types of enrichment materials after the two different fasting times (0 and 12 h). Different superscripts indicate significant differences due to treatment ($P < 0.05$).

3.3. Skin Lesions

As illustrated in Figure 5, the total number of lesions significantly increased over time for all enrichment treatments (F (5, 460) = 131.78, $P < 0.0001$). The time*treatment interaction was also significant. At the first measurement (67 days old, before provision of enrichments), a significantly higher count of lesions was found for the pigs in the paper and straw enrichments, compared to those in the chain or wood treatment (F (15, 460) = 4.42, $P = 0.049$). At 156 days old, the pigs in the paper treatment presented a significantly higher count of body lesions compared to the other three treatments (F (15,460) = 4.42, $P = 0.0027$). At the last evaluation (167 days old), the pigs in the paper treatment showed a significantly higher number of lesions compared with the pigs in the chain and wood treatments which were higher than the number of lesions in the pigs who were offered straw (F (15, 460) = 4.42, $P = 0.0099$).

Concerning the number of lesions, a significant increase was observed in all the specified areas of the body over time ($P < 0.0001$), and the interaction of time*treatment was also significant for some of the areas, as follows. In the rear region (the tail was included in this study), pigs in the paper treatment presented a significantly higher count on the first basal measurement, compared to those pigs in the chain, wood and straw treatments (Table 3, F (15, 460) = 3.71, $P = 0.021$). At the final evaluation, the pigs in the chain and paper treatments showed a significantly higher count of rear lesions compared to those in the wood and straw treatments (F (15, 460) = 3.71, $P = 0.021$). At the first measurement, for the front region, the pigs in the straw treatment had a higher count of lesions compared to those in the other three treatments (F (15, 460) = 2.98, $P = 0.07$). In contrast, at the final measurement, the pigs in the straw treatment showed a significantly lower count compared to those in the other three treatments (F (15, 460) = 2.98, $P = 0.032$). For the ears, side and leg regions, the interaction time*treatment was not significant. No severe tail lesions, i.e., with fresh blood, were observed in any of the treatments at any time.

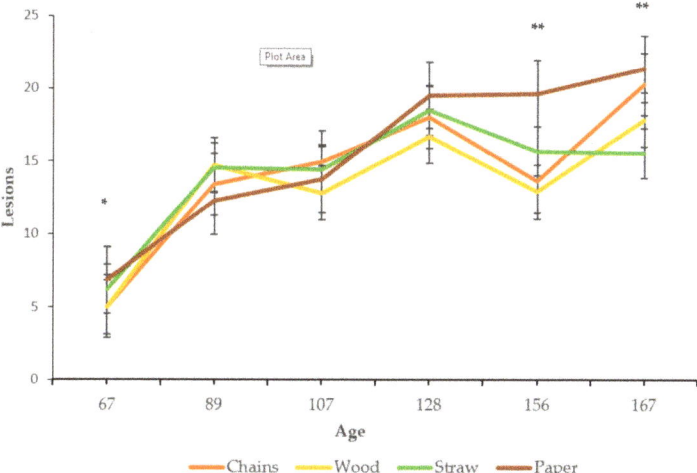

Figure 5. Total number of lesions (counts and standard errors) for the four types of enrichment material over time. * $P < 0.05$; ** $P < 0.01$ (see text for specific differences).

Table 3. Counts of lesions (mean and standard error) in the rear and front regions before the provision of the enrichment treatments (67 days old) and at the final evaluation (167 days old).

	Enrichment Material Treatment				
	Chain	Wood	Straw	Paper	P Treatment * Time [1]
Rear Lesions 67	0.62 [a] ± 0.41	1.08 [a] ± 0.89	0.63 [a] ± 1.57	1.83 [b] ± 1.22	
Rear Lesions 167	7.20 [d] ± 2.90	6.00 [c] ± 1.85	5.66 [c] ± 2.66	7.17 [d] ± 2.62	*
Front Lesions 67	1.87 [a] ± 0.70	1.71 [a] ± 0.56	2.29 [b] ± 1.83	1.79 [a] ± 0.56	
Front Lesions 167	6.2 [d] ± 4.9	5.25 [d] ± 3.35	3.95 [c] ± 3.01	5.62 [d] ± 4.48	*

[1] Different superscripts indicate significant differences due to treatment*time interaction.

3.4. Weight and Productivity Data

The enrichment treatments did not significantly affect any of the productivity parameters that were evaluated (weight, average daily gain, daily feed intake and feed conversion ratio, Table 4). The weight of the pigs increased significantly over time (F (3, 36) = 282.63, $P < 0.0001$), and at the same rate for all four enrichment treatments. Overall the average daily gain (ADG), daily feed intake (DFI) or feed conversion ratio (FCR) were not significantly affected by the enrichment treatments (the average for all treatments was 0.89 ± 0.04 kg/day; 1.96 ± 0.11 kg/day and 2.21 ± 0.06, respectively). The ADG and DFI were significantly lower when the first measurement was taken (at between 68 and 90 days old), compared with the other three measurements (F(ADG) (3,36) = 18.74, F(DFI) (3,36) = 262.83, $P < 0.0001$). The feed conversion ratio was significantly lower at the first two measurements (68–90 and 91–107 days old), compared with the third and fourth measurements (F (3,36) = 103.87, $P < 0.0001$).

Table 4. The Mean (and standard deviations) of Average Daily Gain, Daily Feed Intake and Feed Conversion Ratio for the four types of enrichment material at different periods (1 to 4) and of the initial Weight and Final Weight.

	Enrichment Material Treatment				P treatment [3]	P time
	Chain	Wood	Straw	Paper		
ADG 1 [1,2]	0.77 [a] ± 0.07	0.77 [a] ± 0.03	0.75 [a] ± 0.04	0.74 [a] ± 0.06		
ADG 2	0.92 [b] ± 0.08	0.91 [b] ± 0.13	0.91 [b] ± 0.15	0.92 [b] ± 0.068	NS	***
ADG 3	0.93 [b] ± 0.11	0.92 [b] ± 0.04	0.91 [b] ± 0.02	0.92 [b] ± 0.05		
ADG 4	0.94 [b] ± 0.07	0.97 [b] ± 0.05	0.98 [b] ± 0.04	0.95 [b] ± 0.09		
DFI 1	1.39 [a] ± 0.08	1.41 [a] ± 0.06	1.44 [a] ± 0.07	1.43 [a] ± 0.02		
DFI 2	1.72 [b] ± 0.14	1.63 [b] ± 0.037	1.71 [b] ± 0.19	1.76 [b] ± 0.15	NS	***
DFI 3	2.14 [b] ± 0.19	2.21 [b] ± 0.16	2.18 [b] ± 0.16	2.15 [b] ± 0.11		
DFC 4	2.45 [b] ± 0.19	2.51 [b] ± 0.15	2.56 [b] ± 0.15	2.48 [b] ± 0.13		
FCR 1	1.80 [a] ± 0.07	1.83 [a] ± 0.03	1.90 [a] ± 0.16	1.93 [a] ± 0.13		
FCR 2	1.87 [a] ± 0.12	1.86 [a] ± 0.22	1.90 [a] ± 0.11	1.94 [a] ± 0.12	NS	***
FCR 3	2.32 [b] ± 0.15	2.41 [b] ± 0.14	2.40 [b] ± 0.19	2.34 [b] ± 0.07		
FCR 4	2.62 [b] ± 0.22	2.59 [b] ± 0.24	2.61 [b] ± 0.10	2.61 [b] ± 0.24		
IW	25.60 [b] ± 1.61	27.14 [b] ± 1.71	27.04 [b] ± 1.54	26.20 [b] ± 0.48		
FW	105.94 [b] ± 6.26	107.96 [b] ± 4.12	106.43 [b] ± 4.63	106.16 [b] ± 4.02	NS	***

[1] ADG: Average Daily Gain (in kg/day); DFC: Daily Feed Intake (in kg/day); FCR: Feed Conversion Ratio; IW = weight at 68 days of age (in kg); FW = weight at 157 days of age in kg. [2] Periods (in days of age) = 1: 68-90; 2: 91-107; 3: 108-129; 4: 130-157. [3] Means with different superscripts present significant differences due to treatment (row comparisons) or due to time (column comparisons). ***: $P < 0.001$; NS = not significant. Interaction treatment * time = NS.

3.5. Carcass and Meat Quality Parameters

No significant effects from the enrichment treatments were found in any of the carcasses or meat quality parameters (Table 5, provides a summary of all the parameters evaluated). A significant effect of the farm fasting was found for the carcass weight, where those pigs who were fasted for 12 h had a lower carcass weight than those not fasted ($F(1, 87) = 7.17$, $P = 0.0089$).

Table 5. Means (and standard deviations) of carcass and meat quality parameters for the four types of enrichment material after implementing two different fasting times (0 and 12 h).

	Enrichment Material Treatments				P Fasting [2]
	Chain	Wood	Straw	Paper	
Carcass weight 0 h (kg) [1]	88.98 [a] ± 2.83	87.61 [a] ± 2.72	88.46 [a] ± 2.71	87.83 [a] ± 2.71	*
Carcass Weight 12 h (Kg)	81.36 [b] ± 2.71	85.13 [b] ± 2.71	82.73 [b] ± 2.71	82.99 [b] ± 2.72	
Carcass yield (%) [2]	74.50 ± 0.51	74.71 ± 0.44	74.49 ± 0.48	74.09 ± 0.49	NS
Carcass Lean percentage (%)	59.05 ± 0.68	58.79 ± 0.65	58.46 ± 0.64	59.27 ± 0.65	NS
pHu SM	5.48 ± 0.02	5.50 ± 0.09	5.50 ± 0.019	5.52 ± 0.019	NS
ECu SM	7.56 ± 0.3	8.03 ± 0.28	8.07 ± 0.27	7.88 ± 0.28	NS
L* SM	49.57 ± 0.61	50.27 ± 0.58	49.79 ± 0.57	49.59 ± 0.58	NS
a* SM	6.15 ± 0.36	5.74 ± 0.34	6.13 ± 0.34	6.31 ± 0.34	NS
b* SM	0.12 ± 0.1	−0.19 ± 0.1	0.08 ± 0.1	−0.07 ± 0.1	NS

[1] Fasting times = 0 or 12 h. [2] Carcass lean percentage obtained using the Fat-O-Meat'er; pHu SM: muscle pH at *Semimembranosus* (SM) 24 h p.m.; ECu SM=electrical conductivity measured using the *Semimembranosus*; L*, a*, b*: Luminosity, redness and yellowness.[3] * $P < 0.05$; the Means for carcass weight with different superscripts are significantly different due to fasting (column comparisons). No other significant effects were observed either for enrichment material treatment (row comparisons) or for interaction Enrichment*Fasting.

4. Discussion

The aim of this study was to evaluate the effects of four different types of enrichment materials, feasible for implementation in Spanish pig housing and climate conditions, on different behavioural, welfare, productivity and meat quality indicators. The objective was to evaluate whether these different enrichment materials met the criteria of effectiveness or indicated the possibilities for success as suggested by Van der Weerd and Day and as follows [3]: (1) the enrichment material should increase species-specific behaviour, (2) it should maintain or improve levels of health, (3) it should improve the economics of the production system, and (4) it should be practical to employ. The different parameters and indicators evaluated in this study are discussed with a focus on which of the four criteria they can mostly be associated with. The authors acknowledge that they did not exhaustively investigate all the parameters that could be included in the different criteria, particularly regarding the economic impact. Therefore, criteria three and four are areas that require more research in the future.

4.1. Behavioural Observations: Increase or Allowance in Species-Specific Behaviour

One of the main objectives for providing enrichment materials is to enhance the possibility of pigs performing their exploratory and rooting behaviour. It has been said that pigs reared with environmental enrichment are more stimulated [3]. One of the expected behavioural outcomes of this higher stimulation is an increase in exploratory behaviour, which has been reported more in pigs in enriched environments compared to those kept in barren ones [13,27]. Moreover, pigs have been said to have a preference for chewable, destructible, rootable and deformable materials [6], and therefore, materials meeting those characteristics are considered to better satisfy the pigs' exploratory behavioural needs. The EU Recommendation 2016 [21] classifies such materials as optimal, suboptimal or marginal based on the way they are presented and their nature (i.e., how edible, manipulable, investigable and chewable they are). The materials chosen for this study were, in accordance with the EU Recommendation 2016, marginal (chain), suboptimal (straw and paper) or suboptimal to marginal (wood attached to a chain, since classification of wood depends on its softness). The results found in this study support that, overall, straw is the material that better meets the criteria of enhancing pigs' exploratory behaviour, which is in keeping with the findings of previous studies and expert opinions [3,17]. Both in the focal and scan observations, the pigs were found to interact more with straw in the rack, when compared to both chain (in all observations of periods two and three) and wood (in all focal observations and scans of period two). It was not possible to statistically evaluate the difference with paper, since the pigs took a mean time of 10 min to destroy, manipulate and/or chew the paper that was given to them twice a day. Therefore, almost no paper was left when the observations were carried out, but as this finding was also considered a result, it was decided to adhere to the initial design of the experiment. Paper certainly met the criteria of being manipulable, investigable and chewable, but it only enhanced the exploratory behaviour of the pigs for a short time, and the practicalities of that are discussed further in Section 4.3.2.

The findings of this study also confirm that, somehow, properly or sufficiently manipulative materials help to enhance the exploratory behaviour of pigs directed towards enrichment, and there was a significant increase in the percentage of pen interactions found for the paper treatment (i.e., the treatment with low time availability of enrichment), compared with the other three. This is in keeping with findings from previous studies, which report that pigs in barren environments channel their activities towards pen fixtures [3,27]. However, pigs in the paper treatment tended to be more active and eat more in certain periods compared to those in some of the other treatments, but with a less clear pattern than for interactions with the pen. Moreover, lying behaviour did not increase in the pigs in the paper treatment as a consequence of not being able to interact with the enrichment materials during the observations. Therefore, the scan observation results seem to suggest that when sufficient enrichment is not available, pigs may target a good part of their exploratory needs towards the pen structures. On the other hand, in the focal observations, the pigs in the paper treatment were also found to have a higher incidence of redirected behaviour towards other pigs. This result also supports previous findings that pigs reared in enriched environments show less redirected behaviour such as tail biting, ear chewing or belly-nosing, compared to those kept in barren conditions [28,29]. As previously mentioned, the exploratory behaviour of the pigs is considered out of concern for their welfare, and, consequently, pigs kept in barren conditions seem more likely to not only present with increased redirected behaviour, but also other abnormal behaviour [30].

The lowest significant percentages or counts of interaction with enrichment material were found for chains. Chains are often considered to be 'inappropriate' as an enrichment material, at least when they are not combined with other materials [4,21]. A recently published paper of expert opinions on whether indestructible objects are acceptable as adequate enrichment materials suggests that: 'improving the short metal chain by making it longer, so that it reaches floor level and providing more chain ends, would significantly and considerably improve pig welfare, almost (but not quite) sufficiently to reach the experts' cut-off point for acceptability' [31]. The chain used in this experiment was not like that suggested by these experts, and the findings were in keeping with previous reports that showed how chains produce less exploratory behaviour compared with other more destructible materials [4,19,21]. Another important observation regarding chain is the fact that the percentage of pig interactions decreased over time to a greater extent than with straw and wood. With straw, and to a lesser extent wood, the pigs were found to interact more with the enrichment in the second period and, afterwards. However, this use decreased in the third period to the levels found in the first month, and for chain treatment the decrease was steady over time. A sense of novelty has also been reported as an important attribute for enhancing the effectiveness of enrichment materials [4,19]. This study was not designed to properly evaluate the effect of novelty itself, however the results support previous findings that the pigs' interest in most of the enrichment materials decreased over time and was more pronounced for the chain treatment [19].

4.2. Physiological Parameters and Skin Lesions: Maintenance/Enhancement of Health

To discuss the effects of the four types of enrichment material on health, the indicators chosen relate not only to the immune system (neutrophil/lymphocyte ratio), but also to stress response (CgA, Lactate), and skin lesions. Moreover, the incidence of disease was also recorded as part of normal farm routines. In this sense, pigs were not affected by any pathology or disease outbreak, except for one pig treated for mild lameness. It has been suggested that when enrichment is provided as bedding, especially straw, it may have a negative impact on health, because it can harbour pathogens and increase the levels of dust in housing systems [See 3, for a review]. However, in this study the enrichment was provided as point source objects and, therefore, small differences between the different types of material were expected in relation to the risk of disease dissemination. There were no differences found in the neutrophil/lymphocyte ratio between treatments, only a significant increase over time in all treatments. Several studies have shown that increasing corticosteroids because of stress results in a redistribution of white blood cells involved in the immunological response, such as an increase in neutrophils (N) and a decrease in lymphocytes (L) [32]. For this reason, an increase in the ratio N:L has been suggested as a potential indicator of acute and even chronic stress [32]. On the other hand, previous studies report a higher N:L ratio in weaned pigs compared to fattening pigs at different ages [33]. Therefore, the increase in the N:L ratio over time found in this study could be interpreted as an indication that stress levels increased equally for all the enrichment treatments throughout the experiment and was perhaps associated with other factors such as lower space availability due to growth or higher external temperatures at the end of the experiment. This possible explanation should be interpreted with caution, because the second blood sample was obtained after exsanguination. Unlike the hormonal response to stress, the initial leukocyte response begins over a time span of hours to days depending on factors such as intensity of the stressor [32], but the effects of transport and lairage cannot be discarded, since it was not possible to obtain a baseline level before transportation took place.

CgA is an acidic soluble protein stored and co-released with catecholamines to the blood from the vesicles of the adrenal medulla, the sympathetic nerve endings ad neuroendocrine tissues [24]. Salivary CgA has been described as a good acute stress indicator in pigs [23,24], and in recently published papers there was a hypothesis that it could also be a potential indicator of chronic stress [23]. In this study, the levels of CgA found were similar to those previously reported [23,24], and no significant differences were observed between enrichment treatments in absolute values. However, the CgA levels increased in the second sample when compared with the basal levels and tended to be higher for those pigs in the paper treatment. This is partly in keeping with previous findings [23], in which a significant increase in CgA levels was observed in pigs kept in barren environments compared to those offered a combination of three enrichment materials (rope, sawdust and ball). Some authors [23] suggest that an increase in CgA over time for those pigs not offered enrichment materials could indicate chronic stress. In this experiment, the pigs in the paper treatment spent little time with the enrichment material, similar to those pigs raised with no enrichment in the previous study. However, at the end of the study the increase in the CgA ratio did not differ between treatments, indicating that perhaps the pigs in the paper treatment had adapted physiologically to their conditions. Further experiments are required to ascertain the role of a lack of sufficient and/or appropriate enrichment materials on CgA as an indicator of chronic stress response.

The pigs in this experiment were also subjected to different pre-slaughter fasting times (0 and 12 h). The lack of on farm fasting has been described as a stressor [34] and was used to reveal whether pigs in any of the enrichment treatments had a greater capacity to cope with it. An interaction between fasting time and enrichment treatment was found for CgA, this being the pigs in the straw treatment who seemed to cope better with the more stressful situation of the lack of on farm fasting. There is limited research into the effects of providing enrichment material and the responses to different pre-slaughter conditions, and the results are inconsistent. Geverink et al. [35] found a rise in the pig's cortisol levels due to their transportation being enriched with straw bedding. However, it was difficult to compare the results with those for pigs kept in barren environments because straw was omitted during the hours prior to transportation, and apparently, this led to high baseline cortisol levels. In this study, the enrichment materials were present until all the pigs were loaded inside the truck. This also concurs with the higher lactate levels found in the pigs in the chain treatment compared with those in the straw treatment. Increased lactate levels after stressors such as transportation are interpreted as an indicator of how the animals cope with physical stress, since muscular exercise causes a high demand for oxygen and anaerobic glycolysis [34]. It has been suggested that the signalling pathways of exercise-trained muscles repeatedly exposed to glycogen degradation and resynthesis do adapt, and, thus, a lower increase in lactate levels would be expected from more active pigs. A hypothesis for the lower levels of lactate found in the pigs in the paper and straw treatments in this experiment would be that overall, they presented with higher activity behaviours. However, further research is required, with more samples over time, to clarify the effects of different types of enrichment materials on their exercise inducing capacity and how this influences muscle adaptation to other physical stressors.

Body lesions are included as one of the parameters assessed for the criteria of 'good health' in the Welfare Quality® assessment protocol and are said to be a proxy indicator of aggression [25]. An interaction between time and enrichment treatment was found for the total number of lesions, and for those in the front and rear region (with tail included in this study). Whereas the baseline level for pigs in the straw treatment was higher, at the end of the study there was a greater number of skin lesions for pigs in the chain and paper treatments. This result can be partly associated with the social behaviour and activity observed, which was found to be higher in the paper treatment at the end of the study and when the pigs were provided with straw at the outset. However, this pattern was not associated with the pigs in the chain treatment. Controversial results have been reported in previous studies regarding the effects of different types of enrichment, including aggression and skin lesions. Some investigations describe reduced levels of fighting and aggression in enriched environments whereas others fail to demonstrate this [see 3, for a review]. Confounding factors such as differences in activity, breed and the baseline aggressiveness of pigs have been suggested as potential explanations for the different results. Our findings of a higher incidence of lesions in the paper treatment supports the assumption that deficient or insufficient enrichment materials can enhance competition and, in turn, increase aggression. This is supported by the higher levels of negative social behaviour seen in the focal observations. Moreover, it can be speculated that the higher number of lesions on the rear region of the pigs in the paper treatment is associated with the higher percentage of redirected behaviour, although a more precise scoring system for tail damage is necessary to ascertain this. Our findings are also in agreement with previous studies reporting a reduction in aggression and skin lesions when pigs are offered straw [3,14,17]. However, it is also important to comment on other factors that are common in experimental conditions, such as the lower number of pigs per pen, the positive human-animal relationship, the reduced mixing practices and the high environmental standards when compared to commercial conditions. All of these factors could have influenced the absence of severe lesions and the lack of greater differences between the enrichment treatments in this study.

4.3. Areas for Future Research

4.3.1. Productivity Indicators and Carcass and Meat Quality: The Economics of the Production System

In this study, no differences were found in either the performance parameters or the quality of the carcass and meat between the four types of enrichment materials. Some objects like chains, balls, metal bars, nutritious objects or cloth strips have been reported to have no influence on productivity in grower finishing pigs, which is in keeping with our results [3,36]. On the contrary, some other investigations have found better growth gains, feed intakes or feed conversion ratios, when providing point source enrichment-objects [3,13,15]. However, in most of these studies, it is important to note that the control pigs were kept in barren environments and the other pigs were provided with at least one type of enrichment, whereas in this study all the pigs were offered enrichment. Moreover, some studies have attributed the better performances found for straw bedded systems to the higher levels of activity and exploratory behaviour of the pigs which leads to higher feed intakes. The increase in the levels of exploration by the pigs with straw as bedding can be up to twenty times higher than those with objects [3]. If the underlying cause of better performance is its association with increased activity and exploration, the results obtained in this study could be explained by the increase in activity for pigs in the paper and straw treatment not being sufficient enough to trigger an increase in feed intake.

On the other hand, there were no major differences found in this study between the different enrichment materials when it came to carcass and meat quality, which concurs with previous investigations [35,37]. However, other studies have reported some effects of enriched environments such as a better water-binding capacity and lower shear force, but with no differences in pH values as found in this study [27]. Previous studies also reveal that the ease of handling and behaviour in a novel environment can be influenced by the environmental enrichment [16,35]. As previously discussed, the pigs provided with straw showed significantly lower levels of CgA at slaughter than those offered chain, but none of the carcass or meat quality parameters were affected. It remains to be discovered whether the fasting treatments were sufficient or insufficient as pre-slaughter stressors and affect meat quality, or whether the source objects used can be expected to improve meat quality.

Regarding the cost of the treatments, straw in a rack requires the highest investment and maintenance cost. It is estimated that a mean quantity of 30 g per day per pig of straw was used. This is higher than the quantity of 8.3 g per day per pig reported in a recent study for chopped straw [38]. The same author reports higher amounts in previous trials, which agrees with the results of this study [19]. The differences may be due to the length of straw provided (non-chopped was used in this study) or to the design of the rack system. Therefore, a reduction in the amount of straw and, thus, in the costs could be achieved by considering these aspects. A proper economic evaluation, considering the inputs and outputs is an important area for future research.

4.3.2. Practicalities to Employ

The chain enrichment was chosen for the control group in this study (chain is not appropriate on its own) or because if proven to work, a combination could be easily implementable on Spanish pig farms. Of the four materials tested, paper seemed to trigger pigs' interest, but only for a short time and providing paper more often would be time consuming for farmers. Another concern in relation to newspaper regards the potential toxicity of the ink. As paper is commonly used in some farms, it was used in this study to obtain more information and there appeared to be no detrimental effect on productivity. Chopped and pelleted newspaper has previously been considered safe as animal bedding [39]. Moreover, previous studies did not find any detrimental effect on pre-weaning mortality and weaning weight when using newspaper as enrichment material for piglets [40]. However, it must be pointed out that there is still insufficient information on whether ingesting newspaper is a risk to pigs. This depends on the type of ink used as some do emit volatile compounds such as toluene. Therefore, even if the practical issues previously mentioned can be overcome, more information is required before recommending the use of paper including that which uses new soy and water-based inks. The wood provided was renewed, and softer wood is recommended due to its manipulability and interest to pigs. Finally, the straw in a rack was found to be the most interesting for pigs and practical. The people topping it up regularly said it increased labour time to approximately 15 min a day. However, a major constraint of using straw was highlighted when the slatted floor was lifted at the end of the experiment. A considerable amount of straw was found mixed with the slurry, creating a risk of blockage in the system. Therefore, it is strongly recommended to improve the design of the rack by adding a tray underneath to prevent this problem, or to consider other straw lengths.

5. Conclusions

Straw in a rack is the material found to enhance more species-specific behaviour, such as exploratory behaviour towards the enrichment. In addition, the pigs provided with straw presented with the lowest percentage of redirected behaviour. Chains presented a more pronounced reduction in pigs' interest over time when compared with straw. Paper was only available for a short period after it was provided, and the percentage of redirected behaviour, either towards the pen or to other pen-mates as well the values of CgA in saliva tended to be higher. No major health problems could be attributed to any of the enrichment treatments. Pigs offered straw presented with a lower increase in CgA levels when subjected to a more stressful pre-slaughter treatment (no fasting) when compared with the pigs provided with chains and paper, potentially indicating a greater capacity to cope with this stressful condition. Lactate levels were found to be lower in the pigs in the straw and paper treatments, the ones in which pigs presented with higher levels of on farm activity, which may have allowed them to cope better with physical stress. Skin lesions were lower for the pigs in the straw treatment compared to pigs offered paper, who also had a higher percentage of negative social behaviour. The cost of providing any of the enrichments in this study was not that expensive, although the straw in a rack had a higher labour cost and management expenses. In this study, the economic benefits of providing any of the enrichment materials tested would have to be linked to the potential advantages for animal welfare, because no positive effects on performance or meat quality were found. To summarise, straw in a rack proved to be the best option from an animal welfare point of view. However, some design aspects need to be improved, such as adding a tray to collect the spare straw, to avoid the risk of blocking the slurry systems. It is important to note that the overall welfare status of the pigs in these experimental conditions was good which may have balanced out any differences between the enrichments provided.

Author Contributions: E.F. conceived, designed, conducted the investigation, the statistical analysis (formal analysis) and contributed to the draft of the manuscript. M.M.R. and R.V. collected the experimental data, performed preliminary analysis and M.M.R. contributed to the draft of the manuscript. D.E. and J.J.C. validated the methodology of CgA analysis and contributed to the draft of the manuscript. X.M. and A.V. contributed to the draft of the manuscript. E.F. and A.V. were in charge of funding acquisition, project administration and supervision.

Funding: This research was funded by "Ministerio de Economía y Competitividad" (MINECO) under grant agreement number AGL2015-68373-C2-2-R. Escribano was supported by "Juan de la Cierva Formación" grant of the" Ministerio de Economía y Competitividad (MINECO)" (FJCI-2015-24662), Spain.

Acknowledgments: We would like to thank the technical staff from IRTA Monells Albert Fontquerna, Carlos Millán, Albert Brun and Albert Rossell, for their invaluable technical assistance. Helpful comments on interpretation of the results were given by Marina Gispert, Font-i-Furnols and Dalmau.

Conflicts of Interest: The authors have no conflict of interest.

References

1. Newberry, R.C. Environmental enrichment: Increasing the biological relevance of captive environments. *Appl. Anim. Behav. Sci.* **1995**, *44*, 229–243. [CrossRef]
2. Stolba, A.; Wood-Gush, D.G. The behaviour of pigs in a semi-natural environment. *Anim. Prod.* **1989**, *48*, 419–425. [CrossRef]
3. Van de Weerd, H.A.; Day, J.E.L. A review of environmental enrichment for pigs housed in intensive housing systems. *Appl. Anim. Behav. Sci.* **2009**, *116*, 1–20. [CrossRef]
4. European Food Safety Authority (EFSA). Scientific Opinion of the Panel on Animal Health and Welfare on a request from Commission—The risks associated with tail-biting in pigs and possible means to reduce the need for tail-docking considering the different housing and husbandry systems. *EFSA J.* **2007**, *611*, 1–13.
5. European Food Safety Authority (EFSA). Scientific Opinion concerning a Multifactorial approach on the use of animal and non-animal-based measures to assess the welfare of pigs. *EFSA J.* **2014**, *12*, 3702.
6. Bolhuis, J.E.; Schouten, W.G.P.; Schrama, J.W.; Wiegant, V.M. Behavioural development of pigs with 349 different coping characteristics in barren and substrate-enriched housing conditions. *App. Anim. Behav. Sci.* **2005**, *93*, 213–228. [CrossRef]
7. Schroder-Petersen, D.L.; Simonsen, H.B. Tail biting in pigs. *Veterin. J.* **2001**, *162*, 196–210. [CrossRef] [PubMed]
8. Taylor, N.R.; Main, D.C.J.; Mendl, M.; Edwards, S.A. Tail-biting A new perspective. *Veterin. J.* **2010**, *186*, 137–147. [CrossRef] [PubMed]
9. D'Eath, R.B.; Arnott, G.; Turner, S.P.; Jensen, T.; Lahrmann, H.P.; Busch, M.E.; Niemi, J.K.; Lawrence, A.B.; Sandoe, P. Injurious tail biting in pigs: how can it be controlled in existing systems without tail docking? *Animals* **2014**, *8*, 1479–1497. [CrossRef]
10. European Union. *Council Directive 2008/120/EC of 18 December 2008 Laying Down Minimum Standards for the Protection of Pigs (Amending 91/630/EEC)*. 2009; Official Journal of the European Union: Brussels, Belgium, 2009; Volume 7, pp. 5–13.
11. De Briyne, N.; Berg, C.; Blaha, T.; Palzer, A.; Temple, D. Phasing out Pig Tail Docking in the EU—Present State, Challenges and Possibilities. *Porc. Heal. Manag.* **2018**, *4*, 1–9. [CrossRef]
12. D'Eath, R.B.; Niemi, J.K.; Ahmadi, B.V.; Rutherford, K.M.D.; Ison, S.H.; Turner, S.P.; Anker, H.T.; Jensen, T.; Busch, M.E.; Jensen, K.K.; et al. Why are most EU pigs tail docked? Economic and ethical analysis of four pig housing and management scenarios in the light of EU legislation and animal welfare outcomes. *Animals* **2016**, *10*, 687–699. [CrossRef] [PubMed]
13. Casal, N.; Manteca, X.; Dalmau, A.; Fàbrega, E. Influence of enrichment material and herbal compounds in the behaviour and performance of growing pigs. *Appl. Anim. Behav. Sci.* **2017**, *195*, 38–43. [CrossRef]
14. Day, J.E.L.; Van de Weerd, H.A.; Edwards, S.A. The effect of varying lengths of straw bedding on the behaviour of growing pigs. *Appl. Anim. Behav. Sci.* **2008**, *109*, 249–260. [CrossRef]
15. Van de Weerd, H.A.; Docking, C.M.; Day, J.E.L.; Breuer, L.K.; Edwards, S.A. Effects of species-relevant environmental enrichment on the behaviour and productivity of finishing pigs. *Appl. Anim. Behav. Sci.* **2006**, *99*, 230–247. [CrossRef]
16. Grimberg-Henrici, C.G.E.; Vermaak, P.; Bolhuis, J.E.; Nordquist, R.E.; van der Staay, F.J. Effects of environmental enrichment on cognitive performance of pigs in a spatial holeboard discrimination task. *Anim. Cogn.* **2016**, *19*, 271–283. [CrossRef] [PubMed]
17. Tuyttens, F.A.M. The importance of straw for pig and cattle welfare: a review. *Appl. Anim. Behav. Sci.* **2005**, *92*, 261–282. [CrossRef]
18. Telkänranta, H.; Bracke, M.; Valros, A. Fresh wood reduces tail and ear biting and increases exploratory behaviour in finishing pigs. *Appl. Anim. Behav. Sci.* **2014**, *161*, 50–59.
19. Courboulay, V. Enrichment materials for fattening pigs: Summary of IFIP trials. *Cahiers l'IFIP* **2014**, *1*, 47–56.
20. Averós, X.; Brossard, L.; Dourmad, J.Y.; de Greef, K.H.; Edge, H.L.; Edwards, S.A.; Meunier-Salaün Alaün, M.C. A meta-analysis of the combined effect of housing and environmental enrichment characteristics on the behaviour and performance of pigs. *Appl. Anim. Behav. Sci.* **2010**, *127*, 73–85. [CrossRef]
21. European Union. *Commission Recommendation (EU) 2016/336 of 8 March on the Application of Council Directive 2008/120/EC Laying Down Minimum Standards for the Protection of Pigs as Regards Measures to Reduce the Need for Tail-Docking*; Official Journal of the European Union: Brussels, Belgium, March 2016.

22. Martin, P.; Bateson, P. *Measuring Behaviour—An Introductory Guide*, 2nd ed.; Cambridge University Press: Cambridge, UK, April 1993.
23. Casal, N.; Manteca, X.; Escribabon, D.; Cerón, J.J.; Fàbrega, E. Effect of environmental enrichment and herbal compound supplementation on phyisiological stress indicators (chromogranin A, cortisol and tumor necrosis factor-α) in growing pigs. *Animals* **2017**, *8*, 1–9.
24. Escribano, D.; Soler, L.; Gutiérrez, A.M.; Martínez-Subiela, S.; Cerón, J.J. Measurement of chromogranin A in porcine saliva: Validation of a time-resolved immunofluorometric assay and evaluation of its application as a marker of acute stress. *Animals* **2013**, *7*, 640–647. [CrossRef]
25. Blokhuis, H.J. *Welfare Quality® Assessment Protocol for Pigs (Sows and Piglets, Growing and Finishing Pigs)*; Report for Welfare Quality®Consortium: Lelystad, The Netherlands, October 2009.
26. Cameron, A.C.; Trivedi, P.K. *Regression Analysis of Count Data*; Cambridge University Press: New York, NY, USA, September 1998.
27. Beattie, V.E.; O'Connell, N.E.; Moss, B.W. Influence of environmental enrichment on the behaviour, performance and meat quality of domestic pigs. *Liv. Prod. Sci.* **2000**, *65*, 71–79. [CrossRef]
28. De Jong, I.C.; Ekkel, E.D.; Van De Burgwal, J.A.; Lambooij, E.; Korte, S.M.; Ruis, M.A.W.; Koolhaas, J.M.; Blokhuis, H.J. Effects of strawbedding on physiological responses to stressors and behavior in growing pigs. *Physiol. Behav.* **1998**, *64*, 303–310. [CrossRef]
29. Jensen, M.B.; Studnitz, M.; Pedersen, L.J. The effect of type of rooting material and space allowance on exploration and abnormal behaviour in growing pigs. *Appl. Anim. Behav. Sci.* **2010**, *123*, 87–92. [CrossRef]
30. Studnitz, M.; Jensen, M.B.; Pedersen, L.J. Why do pigs root and in what will they root? *Appl. Anim. Behav. Sci.* **2007**, *107*, 183–197. [CrossRef]
31. Bracke, M.; Koene, P. Expert opinion on metal chains and other indestructible objects as proper enrichment for intensively-farmed pigs. *PLoS ONE* **2019**, *14*, 1–19. [CrossRef] [PubMed]
32. Davis, A.K.; Maney, D.L.; Maerz, J.C. The use of leukocyte profiles to measure stress in vertebrates: A review for ecologists. *Funct Ecol.* **2008**, *22*, 760–772. [CrossRef]
33. Stull, C.L.; Kachulis, C.J.; Farley, J.L.; Koenig, G.J. The effect of age and teat order on α1-acid glycoprotein, neutrophil-to-lymphocyte ratio, cortisol, and average daily gain in commercial growing pigs. *J. Anim. Sci.* **1999**, *77*, 70–74. [CrossRef]
34. Knowles, T.G.; Warriss, P.D.; Vogel, K. Stress physiology of animals during transport. In *Livestock Handling and Transport*; Grandin, T., Ed.; CABI Publishing: Wallingford, UK, 2014; pp. 399–420.
35. Geverink, N.A.; De Jong, I.C.; Lambooij, E.; Blokhuis, H.J.; Wiegant, V.M. Influence of housing conditions on responses of pigs to preslaughter treatment and consequences for meat quality. *Can. J. Ani. Sci.* **1999**, *79*, 285–291. [CrossRef]
36. Day, J.E.L.; Burfoot, A.; Docking, C.M.; Whittaker, X.; Spoolder, H.A.M.; Edwards, S.A. The effects of prior experience of straw and the level of straw provision on the behaviour of growing pigs. *Appl. Anim. Behav. Sci.* **2002**, *76*, 189–202. [CrossRef]
37. Casal, N.; Fon-i-Furnols, M.; Gispert, M.; Manteca, X. Effect of Environmental Enrichment and Herbal Compounds-Supplemented Diet on Pig Carcass, Meat Quality Traits, and consumers' Acceptability and Preference. *Animals* **2018**, *8*, 118. [CrossRef] [PubMed]
38. Coruboulay, V.; Guingand, N. *Paille ou objects à manipuler: quelle attractivité pour le porc en croissance loge sur caillebotis?* 51èmes Journées de la Recherche Porcine: Paris, France, February 2019.
39. Ward, P.L.; Wohlt, J.E.; Zajac, P.K.; Cooper, K.R. Chemical and physical properties of processed newspaper compared to wheat straw and wood shavings as animal bedding. *J. Dairy Sci.* **2000**, *83*, 359–367. [CrossRef]
40. Telkänranta, H.; Swan, K.; Hirvonen, H.; Valros, A. Chewable materials before weaning reduce tail biting in growing pigs. *Appl. Anim. Behav. Sci.* **2014**, *157*, 14–22.

 © 2019 by the authors. Licensee MDPI, Basel, Switzerland. This article is an open access article distributed under the terms and conditions of the Creative Commons Attribution (CC BY) license (http://creativecommons.org/licenses/by/4.0/).

Article

Effect of Different Environment Enrichments on Behaviour and Social Interactions in Growing Pigs

Lorella Giuliotti *, Maria Novella Benvenuti, Alessandro Giannarelli, Chiara Mariti and Angelo Gazzano

Department of Veterinary Science, University of Pisa, Viale delle Piagge, 2, 56124 Pisa, Italy; novella.benvenuti@unipi.it (M.N.B.); a.giannarelli@studenti.unipi.it (A.G.); chiara.mariti@unipi.it (C.M.); angelo.gazzano@unipi.it (A.G.)
* Correspondence: lorella.giuliotti@unipi.it

Received: 23 January 2019; Accepted: 12 March 2019; Published: 19 March 2019

Simple Summary: Pigs reared under intensive conditions are subjected to environmental stresses such as being unable to express some natural behaviours, like socialisation, exploring and rooting. For this reason, EU legislation requires farmers to employ suitable environmental enrichments in the pens. This study aimed at evaluating how different environmental enrichment tools (wooden logs either hanging or laying and hanging metal chains in pens) affected the behaviour of growing pigs. The results show a reduction in the incidence of aggressive/damaging interactions between animals in the pen where hanging wooden logs were placed. No significant effect on non-aggressive behaviours was noted in any of the investigated conditions.

Abstract: (1) Background: Pigs are active animals that require a suitable environment to be able to express their exploratory behaviour. The aim of the present study was to compare the effects of different environmental enrichments on the behaviour, social interactions, salivary cortisol concentration and body weight of pigs during the growing phase. (2) Methods: The investigation involved 75 pigs divided into three groups. The environmental enrichments were arranged as follows: Hanging metal chains for the control group; hanging metal chains and hanging logs for the second group; hanging metal chains and logs laying on the floor for the third group. Each group was video recorded twice a week for six weeks. The scan sampling technique was used. Salivary cortisol and live body weight were also recorded regularly. Parametric (ANOVA) and non-parametric statistics were used to analyse the data. (3) Results: Hanging logs were found to be more effective than logs laying on the floor at reducing aggression within the group tested, resulting in a more comfortable environment. Salivary cortisol concentration and growth did not show significant differences between the three groups. (4) Conclusions: The use of hanging logs affected some interactive patterns that resulted in decreasing the aggressive episodes of pigs, thereby providing a more comfortable environment.

Keywords: pig; environmental enrichment; behaviour; social interactions

1. Introduction

Pigs have an innate propensity for socialisation, exploration, rooting and chewing behaviours. When individuals are unable to express these behaviours, e.g., in poorly enriched environments, abnormal activities may surface [1].

The main housing solutions adopted in the post-weaning and fattening swine sectors are generally not geared up to satisfy the need for expression of these natural behaviours of the species. A monotonous environment can cause stress and induce stereotypic behaviours and apathy, which can then give rise to extremely dangerous phenomena, such as biting the tails and ears of pen-mates [2].

For this reason, environmental enrichments can improve pig welfare by reducing the incidence and severity of behavioural alterations.

EU legislation [3] requires commercial producers to allow pigs to have permanent access to a sufficient quantity of manipulating material, such as straw, hay, wood, sawdust, mushroom compost or peat. However, specific advice on how to provide this enrichment is lacking. In fact, pig farmers must provide such suitable enrichment materials in a way that is compatible with the animal waste handling system, and takes into consideration the cost implications and the impact on animal health [4,5]. Dangling metal chains are commonly used in intensive European pig farms as a form of enrichment [6]. However, this enrichment is not recommended for long-term use, because it quickly loses its novelty, and pigs lose interest in it [7].

What might be the most effective enrichment to improve pig welfare is still debated. This study aims to compare the effects of different environmental enrichments on the behaviour, social interactions, salivary cortisol concentration and body weight of pigs during the growing phase.

2. Materials and Methods

All procedures and treatments were in compliance with the EU Directive 2001/88/EC [8] and EU Directive 2001/93/EC [3] regarding minimum standards for the protection of pigs. Although chains are not entirely considered an adequate enrichment according to EU legislation, they were included in this paper as a control in accordance with Italian legislation [4], which allows the use of other materials when a risk to the functionality of the system exists.

The study was conducted in a farrow-to-finish herd located in the district of Pisa, Italy, from October to December 2016. Seventy-five Goland hybrid grower pigs of both sexes (females and castrated males) were enrolled in the experiment. All subjects were housed in the same building equipped with concrete-slatted-floors pens (2.4 × 6.1 m) and an automatically controlled natural ventilation system. The animals were checked twice a day, and artificial lighting was provided from sunset to 8 pm. The pigs had free access to water from two nipple drinkers per pen, and wet meals were offered every 3 h from 7 am to 7 pm (5 meals/day in total).

A set of dangling metal chains, usually adopted in the farm as environmental enrichment, were present in each pen. The 75 subjects were divided into three groups of 25 each. These groups were homogeneous in body weight (34.9 ± 2.57 kg), sex (13 females and 12 castrated males) and age (11 weeks). The animals were randomly assigned to the three groups:

- Control (C): only dangling metal chains were offered to the animals.
- Hanging logs (HL): three small logs of wood (30 cm) were hooked to the metal chains and hung at 60 cm from the ground.
- Laying logs (LL): three small logs of woods (30 cm) were placed on the floor. Subjects in this group were thus able to interact with both the logs and the dangling metal chains.

The logs were made of poplar, which was selected thanks to its being suitable for chewing and manipulation, harmless, readily available, economical and easy to fit into the farming routine. The wooden enrichment was introduced at the beginning of the trial; as they deteriorated, the logs in LL were replaced almost every two days while the logs in HL were never replaced.

The trial began after a one-week adaptation period. Videos were recorded twice a week for a total period of six weeks by means of Go-Pro Hero cameras placed in front of each of the three pens to gain a complete view of the subjects. On each day of observation, two 90-min recording sessions took place, one in the morning (around 09:00) and one in the afternoon (around 15:00), far enough from the feeding time to ensure a good level of activity and to minimise confounding due to eating behaviour.

Observations of non-social behaviours and social interactions were conducted blind by the same trained observer throughout the experiment. Behavioural observations were carried out according to the scan sampling technique [9], i.e., 30-s observations every 5 min, and the number of subjects that were performing the specific actions outlined in Table 1 was reported.

Table 1. Definition of the analysed behaviours.

Behaviour	Definition
Standing inactive	Subjects stay and do not exhibit any behaviours.
Laying inactive	Subjects lay motionless in lateral or sternal recumbency.
Eating	Subjects stand in front of the feeder and put their head in contact with the feeding trough.
Drinking	Subjects stand and either touch or play with the nipple drinkers.
Social activity	Subjects chase, shove or scratch pen-mates with their snout, bite ears, feet or tails of pen-mates and perform other aggressive behaviours.
Pen exploration	Subjects move around the pen rooting about the floor.
Enrichments examination	Subjects smell, chew, suck or play with the enrichments.
Log examination	Subjects smell, chew, suck or play with the logs that are either hanging from the chains or laying on the floor

The number and type of social interactions were recorded by applying the behaviour sampling method: the same videos were fast-forwarded and paused at the moment in which two or more pigs came into contact with one another. The social interactions (Table 2) were assembled on the basis of the ethogram outlined by Jensen [10]. Moreover, in order to facilitate data analysis, the social interactions were categorised as either "aggressive/damaging" or "non-aggressive" [11,12].

Table 2. Description of the observed social interactions.

Interaction	Labelling	Definition of the Interaction
Parallel pressing	Aggressive/damaging	The pigs stand side by side and shove one another until one attempts to bite the other's head, neck or flank.
Inverse pressing	Aggressive/damaging	The pigs stand in front of one another and push with their heads against the other's neck or flank.
Head-to-head knock	Aggressive/damaging	A pig uses its head or snout to strike another's head, neck or ears. This action may be followed by a bite.
Head-to-body knock	Aggressive/damaging	A pig uses its head or snout to strike another pig with a quick blow to any body part behind the ears. This action may be followed by a bite.
Ears or tail biting	Aggressive/damaging	A pig chews, sucks or plays with another's ears or tails.
Belly nosing	Non-aggressive	A pig uses its snout to repeatedly and continuously massage the abdominal or groin area of another pig that is laying down.
Nose-to-nose	Non-aggressive	A pig places its snout near the head, ears or nose of another pig. A short physical contact may be established.
Nose-to-body	Non-aggressive	A pig places its snout close to the body of another pig, but not in the genital area. A short physical contact may be established.
Anogenital nosing	Non-aggressive	A pig places its snout near the genital area of another. A short physical contact may be established
Withdrawing	Non-aggressive	A pig feels threatened by another, and consequently moves away rapidly while holding its head high and often emitting a shrill cry.

To appraise the evolution of non-social behaviours and social interactions during the trial, the experimental period was divided into three phases of 2 weeks each: Initial (1st and 2nd week), mid-term (3th and 4th week) and final (5th and 6th week).

At the beginning of the trial and every two weeks, salivary cortisol samples were gathered in each pen for a total of four samplings. The samples were collected using large cotton swabs deposited in the pen. These swabs were randomly chewed by the pigs for 30 s and were withdrawn immediately afterwards. The samples were preserved at 4 °C in tubes until they arrived at the Etovet laboratory of the Department of Veterinary Science of the University of Pisa to be stored at −20 °C until the analysis. Every sample was assayed for salivary cortisol using an enzyme immunoassay kit (Diametra® Cortisol Saliva, Spello, Perugia, I).

The body weight of all pigs was measured at the beginning and at the end of the trial. During the experiment, some subjects (three pigs from the HL group, three from the LL group and five from the C group) were removed for illness and the data were promptly updated. Results for non-social behaviours and social interactions were expressed as the percentage of the total number of pigs who performed an activity.

Statistical analyses were performed by the SAS-JMP software [13], as follows:

- ANOVA for the behavioural data; the model included group, period and observation time (morning or afternoon) as variables;
- Wilcoxon nonparametric test for social interactions;
- ANOVA for final body weight and salivary cortisol, using the group as variability factor and initial body weight as covariate only in the former.

A p-value ≤ 0.05 was considered significant.

3. Results and Discussion

3.1. Behavioural Data

Regarding the behavioural differences between the groups (Table 3), statistical analysis uncovered significant differences for "social activity" and "enrichment examination" ($p \leq 0.01$), as well as on "pen exploration" and "log examination" ($p \leq 0.05$).

Table 3. Behavioural observations (%) by group.

Group	C	HL	LL	Standard Error	p Value
Parameter	Mean	Mean	Mean		
Inactive	80.8	76.0	76.7	2.50	0.1688
Active	19.4	24.6	20.6	2.49	0.1447
Standing inactive	0.6	1.0	0.6	0.28	0.3094
Laying inactive	80.2	75.0	76.1	2.57	0.1558
Eating	2.6	2.1	2.0	0.45	0.4363
Drinking	0.4	0.5	0.3	0.90	0.2769
Social activity	1.4 ^{A}	1.2 ^{A}	0.5 ^{B}	0.24	0.0010
Pen exploration	11.9 ^{AB}	14.6 ^{A}	10.2 ^{B}	1.58	0.0413
Enrichments examination	3.0 ^{B}	6.2 ^{A}	7.6 ^{A}	0.94	<0.0001
Log examination	-	6.2	3.3	1.44	0.0524

Different superscript letters in the same row indicate significant differences (B < A). Legend: C (control), HL (hanging logs) and LL (laying logs).

The "pen exploration" trend in C and HL was in accordance with data reported by Beattie et al. [14], whose studies indicated that environmental enrichment increased the time spent in exploratory behaviour. The same authors also reported a positive effect of environmental enrichment on active behaviours, but in our study this effect did not reach statistical significance.

"Social activity" was significantly lower in LL; in fact, the presence of material that could be manipulated seemed to distract the animals from social interactions (both aggressive/damaging and non-aggressive), and this effect was particularly pronounced in the LL group, where the animals could effectively avail themselves of both logs and chains. This result was in accordance with the study of Beattie et al. [14], who affirmed that persistent interactions among pigs represent a redirected impulse of environment manipulation.

"Enrichment examination" was higher in the HL and LL than in the C group, in agreement with Telkänranta et al. [15], who found that pigs prefer to manipulate wooden logs than chains. "Log examination" tended to be higher in HL than in LL. The lower interest of the animals in the LL pen in the wooden logs can be explained by the fact that the logs might be soiled with faeces, becoming unappealing to the animals, as hypothesised by Battini et al. [16].

The most commonly observed behaviours were "laying inactive" followed by "pen exploration" and "eating," in agreement with Cornale et al. [17], who observed that pigs spent most of their time resting on the floor (>50%), exploring the pen and standing at the through.

3.2. Social Interactions

The statistical analysis showed significant differences in social interactions (Table 4) between the groups for the variables "head-to-head knock," "biting ear or tail," "belly nosing," and "aggressive interactions."

Table 4. Observation of social interactions (%) between the three groups.

Group	C		HL		LL		p Value
Parameter	Mean	SE	Mean	SE	Mean	SE	
Aggressive/damaging interactions	46.35 ^{A}	6.65	24.10 ^{B}	3.35	32.73 ^{AB}	4.90	0.0265
Non-aggressive interactions	53.59	6.27	59.06	6.02	48.16	4.88	0.3435
Parallel pressing	6.99	1.6	5.86	1.05	5.02	1.11	0.8488
Inverse pressing	4.82	1.07	2.64	0.58	3.51	0.75	0.2260
Head-to-head knock	22.52 ^{A}	2.65	10.76 ^{B}	1.56	15.99 ^{B}	2.05	0.0006
Head-to-body knock	1.29	0.26	1.59	0.39	0.75	0.24	0.1564
Ears or tail biting	10.73 ^{A}	1.59	3.25 ^{B}	0.73	7.46 ^{A}	1.12	0.0002
Belly nosing	5.27 ^{B}	1.29	10.96 ^{A}	2.07	0.76 ^{C}	0.2	<0.0001
Nose-to-nose	20.77	2.84	21.56	2.24	22.80	2.7	0.7802
Nose-to-body	17.03	1.99	13.45	0.82	16.67	0.64	0.3328
Anogenital nosing	7.87	1.10	8.80	0.76	6.27	0.95	0.0620
Withdrawing	2.65	1.47	4.29	0.65	1.66	0.47	0.1232

Different superscript letters on the same row indicate significant differences (B < A). Legend: C (control), HL (hanging logs) and LL (laying logs). SE = standard error.

"Aggressive/damaging interactions" were higher in the C group than in HL. This difference was mainly caused by the values of the parameters "head-to-head knock" and "ear and tail biting". Similar results were observed in a trial conducted by Cornale et al. [17], in which pigs reared in unenriched pens showed higher rates of tail biting and aggression compared to pigs reared in pen equipped with hanging wooden logs. Other studies detected a reduction in the incidence of aggression among pigs reared in an enriched environment [18,19]. However, there are also studies that show higher levels of aggressive activities in pigs reared in enriched environments [20].

"Belly nosing" was the greatest in HL, lower in C and lower still in LL. The possible cause of belly nosing has not yet been clarified. It has been suggested that one of the principal causes can be the early age of weaning [21]. Therefore, it has been hypothesised that the social environment can also have a profound effect on the incidence of belly nosing [22].

3.3. Behavioural Observation and Social Interactions by Period

Variations in behaviour along the trial are summarised in Table 5. Significant differences were recorded for "log examination" ($p \leq 0.01$) as well as for the variables "drinking" and "pen exploration" ($p \leq 0.05$).

Petersen et al. [23] reported that eating behaviour increased substantially across time concurrently with the age of the animal, while in the present study "eating" remained steady during the trial. "Log examination" followed a decreasing trend in the final period of the trial, possibly due to the loss of interest in the enrichment over time, as found by Van de Weerd et al. [24].

Table 6 reports the time variation of social interactions over time.

Belly nosing showed a decreasing trend during the study. This was in accordance with the findings of Torrey and Widowsk [21], who defined this activity as a transient pattern, representing a redirected suckling behaviour in confined pigs.

"Parallel pressing" was significantly variable during the trial period, while "anogenital nosing" showed a decreasing trend. The later interaction represents a mechanism of mutual recognition [10].

Thus, it is hypothesised that the decrease reflects greater acquaintance and a relatively more stable hierarchy among the pigs in the final period.

Table 5. Behavioural observations by period (%).

Period	Initial	Mid-Term	Final	Standard Error	p Value
Parameter	Mean	Mean	Mean		
Inactive	76.8	75.8	81.0	1.93	0.1299
Active	24.8	24.2	18.8	1.99	0.0700
Standing inactive	0.9	0.7	0.6	0.22	0.7894
Laying inactive	75.9	75.1	80.4	1.99	0.1320
Eating	1.7	2.8	2.1	0.35	0.0944
Drinking	0.3 ^{B}	0.5 ^{A}	0.3 ^{B}	0.07	0.0347
Social activity	1.3	0.9	1.0	0.19	0.2769
Pen exploration	11.8 ^{AB}	14.5 ^{A}	10.4 ^{B}	1.22	0.0562
Enrichment examination	6.9	5.3	4.7	0.72	0.0851
Log examination	9.7 ^{A}	5.5 ^{B}	5.0 ^{B}	0.93	0.0011

Different superscript letters in the same row indicate significant differences (B < A).

Table 6. Social interaction by period (%).

Period	Initial		Mid-Term		Final		p Value
Parameter	Mean	SE	Mean	SE	Mean	SE	
Aggressive/damaging interactions	32.50	4.19	43.89	7.28	26.82	3.64	0.2681
Non-aggressive interactions	57.67	5.11	54.16	5.53	49.01	5.61	0.3643
Parallel pressing	5.58 ^{AB}	0.10	8.74 ^{A}	1.67	3.55 ^{B}	0.76	0.0592
Inverse pressing	2.33	0.46	5.49	1.00	3.15	1.70	0.1321
Head-to-head knock	16.79	2.05	19.26	3.03	13.23	1.59	0.4315
Head-to-body knock	1.13	0.27	1.69	0.40	0.82	0.20	0.4376
Ears or tail biting	6.67	1.07	8.71	1.67	6.07	1.20	0.5496
Belly nosing	9.92 ^{A}	2.33	3.66 ^{B}	1.03	3.42 ^{B}	0.79	0.0049
Nose-to-nose	19.25	1.64	24.90	2.67	20.99	3.15	0.3189
Nose-to-body	15.83	0.85	15.03	1.57	16.30	2.11	0.4766
Anogenitals nosing	10.17 ^{A}	0.76	6.89 ^{B}	0.55	5.88 ^{B}	0.70	0.0013
Withdrawing	2.50	0.61	3.68	0.86	2.42	0.72	0.3206

Different superscript letters on the same row indicate significant differences (B < A).

3.4. Observation during the Day

Table 7 reporting the behavioural changes across the day reveals that all the observed parameters that showed statistical differences.

Table 7. Behavioural observation by time of day (%).

Time of Day Parameter	Morning	Afternoon	Standard Error	p Value
	Mean	Mean		
Inactive	90.2	69.9	2.23	<0.0001
Active	13.0	29.9	1.70	<0.0001
Standing inactive	4.9	1.0	0.25	0.0445
Laying inactive	85.3	68.9	2.30	<0.0001
Eating	1.3	3.1	0.40	<0.0001
Drinking	0.2	0.5	0.08	<0.0001
Social activity	0.6	1.5	0.22	0.0002
Pen exploration	7.2	17.3	1.41	<0.0001
Enrichment examination	3.7	7.5	0.84	<0.0001
Log examination	5.0	8.6	1.10	0.0010

Every behaviour significantly differed across the time of the day, showing greater activity in the afternoon. Consequently, parameters such as "standing inactive" and "laying inactive", representing inactivity, were prevalent in the morning. This general pattern was also recorded by Fraser et al. [25], who observed that pigs showed greater levels of activity in the afternoon rather than in the morning.

Significant differences in social interaction were observed across time of day as reported in Table 8.

Table 8. Social interactions at different timed of the day (%).

Time of the Day	Morning		Afternoon		p Value
Parameter	Mean	SE	Mean	SE	
Aggressive/damaging interactions	16.13	1.58	51.78	4.54	<0.0001
Non-aggressive interactions	30.55	1.84	76.87	3.30	<0.0001
Parallel pressing	1.50	0.25	10.42	0.99	<0.0001
Inverse pressing	0.10	0.17	6.32	0.71	<0.0001
Head-to-head knock	10.20	0.89	22.65	2.08	<0.0001
Head-to-body knock	0.35	0.11	2.07	0.27	<0.0001
Ears or tail biting	3.98	0.58	10.32	0.13	<0.0001
Belly nosing	2.67	0.55	8.66	1.68	0.0033
Nose-to-nose	11.46	0.78	31.97	1.54	<0.0001
Nose-to-body	10.19	0.66	21.25	1.08	<0.0001
Anogenitals nosing	5.78	0.53	9.71	0.57	<0.0001
Withdrawing	0.45	0.12	5.28	0.62	<0.0001

Every interaction was more frequent in the afternoon than in the morning in all groups; these outcomes are in accordance with the findings reported by Ott et al., [26].

3.5. Growing Performance and Cortisol

No significant differences in the final weight among the three groups were found in this study (Table 9). The literature on this issue is controversial: Schaefer et al. [18], Horrell [27] and Beattie et al., [14] found a better growth rate in pigs reared in an enriched environment, whereas Pearce et al. [28] and Blackshaw et al. [29], in accordance with our findings, did not notice any weight gain.

No significant differences in salivary cortisol among the groups and the sampling were found (Tables 9 and 10).

Table 9. Final body weight (kg) and salivary cortisol concentration (ng/mL) of the three groups.

Group	C	HL	LL	Standard Error	p Value
Parameter	Mean	Mean	Mean		
Final body weight	56.06	55.65	59.14	4.99	0.8646
Salivary cortisol	4.92	5.97	4.65	0.94	0.6032

Legends: C (control), HL (hanging logs) and LL (laying logs).

Table 10. Salivary cortisol concentration (ng/mL) during the trial.

Sampling	1	2	3	4	Standard Error	p Value
Parameter	Mean	Mean	Mean	Mean		
Salivary cortisol	6.48	4.86	3.57	5.82	1.09	0.3433

These results confirm the findings of Cornale et al. [17], who reported that the use of hanging wooden logs did not result in significant differences in faecal corticosteroid levels. De Jong et al. [30] found levels in cortisol concentration near to 8 and 6 ng/mL in the 15-week-old pigs reared in enriched and barren environments, respectively. Comparable values were also observed by Smulders et al. [31] in 14 to 20-week-old piglets.

4. Conclusions

This study of swine behaviour in response to enrichment yielded interesting results. In detail, our data suggested that the adoption of hanging wooden logs would allow a reduction in the incidence of aggressive/damaging interactions among the animals. At the same time, there was no significant effect on non-aggressive interactions in any of the investigated conditions.

Regarding the levels of activity throughout the day, as expected, active behaviours and interactions were more frequent in the afternoon than in the morning regardless of the kind of enrichment. The proposed environmental enrichments did not induce significant variation in the growth rate and salivary cortisol.

The use of wooden logs (both hanging and laying) showed a decreasing trend during the trial, possibly due to a decline in interest of the animals towards the items. Overall, the pigs interacted more often with the hanging logs, probably because they were not soiled with faeces like the lying logs.

Although the implementation of hanging wooden logs brought positive results, further investigation is necessary in order to verify whether the interest of the animals can be maintained across the time by modifying the enrichment configuration. Moreover, to highlight the differences due to treatments, the replication of the group should be considered.

Finally, in compliance with the recommendation of the EU legislation and in light of our results, hanging metal chains should be replaced with materials that do not damage the functionality of the waste system.

Author Contributions: Conceptualization, L.G.; Methodology, L.G., M.N.B. and C.M.; Formal Analysis, M.N.B.; Investigation, L.G. and A.G. (Alessandro Giannarelli); Data Curation, L.G. and M.N.B.; Writing—Original Draft Preparation, L.G. and M.N.B.; Writing—Review and Editing, C.M. and A.G. (Angelo Gazzano); Supervision, A.G. (Angelo Gazzano).

Funding: This research received no external funding.

Acknowledgments: The authors wish to thank Beatrice Torracca for her help in the lab, Benedetta Sarno for the English revision and the owner and the staff of the farm.

Conflicts of Interest: The authors declare no conflict of interest.

References

1. Studnitz, M.; Jensen, M.B.; Pedersen, L.J. Why do pigs root and in what will they root? A review on the exploratory behaviour of pigs in relation to environmental enrichment. *Appl. Anim. Behav. Sci.* **2007**, *107*, 183–197. [CrossRef]
2. Day, J.E.L.; Spoolder, H.A.M.; Burfoot, A.; Chamberlain, H.L.; Edwards, S.A. The separate and interactive effects of handling and environmental enrichment on the behaviour and welfare of growing pigs. *Appl. Anim. Behav. Sci.* **2001**, *75*, 177–192. [CrossRef]
3. European Council. *European Council Regulation (EC) 2001/93 of 9 October 2001 Amending Regulation (EC) 91/630 Laying Down Minimum Standards for the Protection of Pigs*; European Council: Hoboken, NJ, USA, 2001.
4. Circular of 2 March 2005 of the Ministry of Health (Procedures for the control of animal welfare in pig farms). Application of Legislative Decree n. 53 of 20 February 2004: Implementation of Directive 2001/93/EC, Establishing Minimal Rules for Pig Protection. Available online: www.normativasanitaria.it/jsp/dettaglio.jsp?id=25268 (accessed on 15 March 2019).
5. Guy, J.H.; Meads, Z.A.; Shiel, R.S.; Edwards, S.A. The effect of combining different environmental enrichment materials on enrichment use by growing pigs. *Appl. Anim. Behav. Sci.* **2013**, *144*, 102–107. [CrossRef]
6. Bracke, M.B.M.; De Lauwere, C.C.; Wind, S.M.M.; Zonerland, J.J. Attitudes of Dutch Pig Farmers Towards Tail Biting and Tail Docking. *J. Agric. Environ. Ethics* **2013**, *26*, 847–868. [CrossRef]
7. Mkwanazi, M.V.; Ncobela, C.P.; Kanengoni, A.T.; Chimonyo, M. Effects of environmental enrichment on behaviour, physiology and performance of pigs—A review. *Asian-Australas J. Anim. Sci.* **2019**, *32*, 1–13. [CrossRef]
8. European Council. *European Council Regulation (EC) 2001/88 of 23 October 2001 Amending Regulation (EC) 91/630 Laying Down Minimum Standards for the Protection of Pigs*; European Council: Hoboken, NJ, USA, 2001.
9. Martin, P.; Bateson, P. *Measuring Behaviour, an Introductory Guide*, 3rd ed.; Cambridge University Press: Cambridge, UK, 2007.
10. Jensen, P. An ethogram of social interaction patterns in group-housed dry sows. *Appl. Anim. Ethol.* **1980**, *6*, 341–351. [CrossRef]

11. Jensen, P. An analysis of agonistic interaction patterns in group-housed dry sows—Aggression regulation through an "avoidance order". *Appl. Anim. Ethol.* **1982**, *9*, 47–61. [CrossRef]
12. Oczak, M.; Viazzi, S.; Ismayilova, G.; Sonoda, L.T.; Roulston, N.; Fels, M.; Bahr, C.; Hartung, J.; Guarino, M.; Berckmans, D.; et al. Classification of aggressive behaviour in pigs by activity index and multilayer feed forward neural network. *Biosyst. Eng.* **2014**, *119*, 89–97. [CrossRef]
13. SAS JMP. *User's Guide, Ver. 5.0*; SAS Inst.: Cary, NC, USA, 2002.
14. Beattie, V.E.; O'Connell, N.E.; Moss, B.W. Influence of environmental enrichment on the behaviour, performance and meat quality of domestic pigs. *Livest. Prod. Sci.* **2000**, *65*, 71–79. [CrossRef]
15. Telkänranta, H.; Brackeb, B.M.; Valrosa, A. Fresh wood reduces tail and ear biting and increases exploratory behaviour in finishing pigs. *Appl. Anim. Behav. Sci.* **2014**, *161*, 51–59. [CrossRef]
16. Battini, M.; Barbieri, S.; Guizzardi, F.; Minero, M.; Canali, E. Effetto di differenti arricchimenti ambientali sul benessere di suini nella fase di ingrasso. *Large Anim. Rev.* **2013**, *19*, 186–190.
17. Cornale, P.; Macchi, E.; Miretti, S.; Renna, M.; Lussiana, C.; Perona, G.; Mimosi, A. Effects of stocking density and environmental enrichment on behavior and fecal corticosteroid levels of pigs under commercial farm conditions. *J. Vet. Behav.* **2015**, *10*, 569–576. [CrossRef]
18. Schaefer, A.L.; Salomons, M.O.; Tong, A.K.W.; Sather, A.P.; Lepage, P. The effect of environment enrichment on aggression in newly weaned pigs. *Appl. Anim. Behav. Sci.* **1990**, *27*, 41–52. [CrossRef]
19. Ishiwata, T.; Uetake, K.; Tanaka, T. Factors affecting agonistic interactions of weanling pigs after grouping in pens with a tire. *Anim. Sci. J.* **2004**, *75*, 71–78. [CrossRef]
20. Bolhuis, J.E.; Schouten, W.G.P.; Schrama, J.W.; Wiegant, V.M. Effects of rearing and housing environment on behaviour and performance of pigs with different coping characteristics. *Appl. Anim. Behav. Sci.* **2006**, *101*, 68–85. [CrossRef]
21. Torrey, S.; Widowski, T.M. Is belly nosing redirected suckling behaviour? *Appl. Anim. Behav. Sci.* **2006**, *101*, 288–304. [CrossRef]
22. Bench, C.J.; Gonyou, H.W. Effect of environmental enrichment at two stages of development on belly nosing in piglets weaned at fourteen days. *J. Anim. Sci.* **2006**, *84*, 3397–3403. [CrossRef] [PubMed]
23. Petersen, V. The development of feeding and investigatory behaviour in free-ranging domestic pigs during their first 18 weeks of life. *Appl. Anim. Behav. Sci.* **1994**, *42*, 87–98. [CrossRef]
24. Van de Weerd, H.A.; Docking, C.M.; Day, J.E.L.; Breuer, K.; Edwards, S.A. Effects of species-relevant environmental enrichment on the behaviour and productivity of finishing pigs. *Appl. Anim. Behav. Sci.* **2006**, *99*, 230–247. [CrossRef]
25. Fraser, D.; Phillips, P.A.; Thompson, B.K.; Tennessen, T. Effect of straw on the behaviour of growing pigs. *Appl. Anim. Behav. Sci.* **1991**, *30*, 307–318. [CrossRef]
26. Ott, S.; Moons, C.B.H.; Kashiha, M.A.; Bahr, C.; Tuyttens, F.A.M.; Berckmans, D.; Niewold, T.A. Automated video analysis of pig activity at pen level highly correlates to human observations of behavioural activities. *Livest. Sci.* **2014**, *160*, 132–137. [CrossRef]
27. Horrell, I. Effects of environmental enrichment on growing pigs. *Anim. Prod.* **1992**, *54*, 483.
28. Pearce, G.P.; Paterson, A.M. The effect of space restriction and provision of toys during rearing on the behaviour, productivity and physiology of male pigs. *Appl. Anim. Behav. Sci.* **1993**, *36*, 11–28. [CrossRef]
29. Blackshaw, J.K.; Thomas, F.J.; Lee, J.A. The effect of a fixed or free toy on the growth rate and aggressive behaviour of weaned pigs and the influence of hierarchy on initial investigation of the toys. *Appl. Anim. Behav. Sci.* **1997**, *53*, 203–212. [CrossRef]
30. De Jong, I.C.; Prelle, I.T.; Burgwal, J.A.; Lambooij, E.; Korte, S.M.; Blokhuis, H.J. Effects of environmental enrichment on behavioural responses to novelty, learning, and memory, and the circadian rhythm in cortisol in growing pigs. *Physiol. Behav.* **2000**, *68*, 571–578. [CrossRef]
31. Smulders, D.; Verbeke, G.; Mormède, P.; Geers, R. Validation of a behavioral observation tool to assess pig welfare. *Physiol. Behav.* **2006**, *89*, 438–447. [CrossRef] [PubMed]

© 2019 by the authors. Licensee MDPI, Basel, Switzerland. This article is an open access article distributed under the terms and conditions of the Creative Commons Attribution (CC BY) license (http://creativecommons.org/licenses/by/4.0/).

Article

Growing Pigs' Interest in Enrichment Objects with Different Characteristics and Cleanliness

Jean-Michel Beaudoin [1,*], Renée Bergeron [2], Nicolas Devillers [3] and Jean-Paul Laforest [1]

1. Département des sciences animales, Université Laval, 2425 rue de l'Agriculture, Québec City, QC G1V 0A6, Canada; jean-paul.laforest@vrrh.ulaval.ca
2. Department of Animal Biosciences, Animal Science and Nutrition, 50 Stone Road East, Guelph, ON N1G 2W1, Canada; rbergero@uoguelph.ca
3. Agriculture and Agri-Food Canada, Sherbrooke Research and Development Centre, 2000 College Street, Sherbrooke, QC J1M 0C8, Canada; nicolas.devillers@canada.ca
* Correspondence: jean-michel.beaudoin.2@ulaval.ca

Received: 22 January 2019; Accepted: 2 March 2019; Published: 8 March 2019

Simple Summary: The modern swine industry is mostly based on an intensive production model that has evolved under economic pressure and has shaped rearing facilities around production optimization rather than the natural needs of pigs. Barren rearing spaces for growing pigs do not allow them to fully express their natural behaviors (e.g., rooting and chewing). This can lead to the emergence of abnormal behaviors, such as tail-biting which causes stress to the animals, and potential financial loss. A simple strategy to allow pigs to express their natural behaviors is to add enrichment objects to the rearing environment. However, pigs tend to lose interest in the objects rapidly. The characteristics of an object, such as the degree of cleanliness and malleability of the material used, can significantly increase its attractiveness. This study first compared seven different objects based on the level of manipulation received from growing pigs. A block of dried wood that was presented on the floor had the longest manipulation time. Secondly, four objects were compared for their level of cleanliness or wear and no differences in manipulation were found between objects that were cleaned or replaced daily and objects that were not cleaned or replaced (for a period of five days).

Abstract: Enrichment objects can be a practical way to provide rooting and chewing material to growing pigs, on which they can express species-specific behaviors. The challenge is to provide enrichment objects that will satisfy pigs' behavioral needs, while being practical and low-cost for the producers. Two trials were conducted to evaluate the effects of object characteristics such as design, location, cleanliness or degree of wear, on pigs' interest over time. The first trial compared seven objects, varying in their design and location, presented individually for five consecutive days to groups of 12 ± 3 (average ± SD) pigs, weighing 61 ± 9.2 kg. The pigs' interest in the objects was evaluated based on the frequency, total duration and mean length of manipulation with the objects. All objects were manipulated at different levels depending on their characteristics. On average, the pigs interacted more frequently ($p < 0.001$) with a chewable object made of three polyurethane balls, spring-mounted and anchored to the floor, and spent more time manipulating a dried wood beam on the floor ($p < 0.05$), which was destructible and chewable, than suspended ropes, plastics and rubber objects, and a plastic ball on the floor. The second trial used two-choice preference tests to compare objects varying in their degree of cleanliness or wear, presented in pairs to growing pigs weighing 47 ± 7 kg and housed in groups of 14 ± 1. Two identical objects were placed simultaneously in a pen over 5 days, and only one of them was cleaned or replaced daily (treatment) while the duplicate was left untouched (control). The results showed no clear preference between control and treatment objects, indicating that short-term maintenance of the objects might be unnecessary.

Keywords: fattening pigs; pig behavior; animal welfare; environmental enrichment

1. Introduction

The welfare of growing pigs has been an increasing concern over recent years and is becoming an important pillar to ensure the sustainability of the swine industry [1]. In many swine-producing countries, pigs are typically raised in barren pens without litter [2]. Such an environment does not allow pigs to fully express their natural species-specific behaviors, which can lead to psychological distress [3]. These pigs become more likely to express abnormal behaviors, such as tail-biting, to cope with the repression of their natural behaviors [3] or as a result of frustration [4]. Tail-biting between pigs can affect health, growth and the welfare of victim pigs and cause significant economic losses for producers [2,5,6]. In European countries, it was estimated that 30% to 70% of farms had tail-biting problems to some extent, with 1% to 5% of pigs presenting tails with lesions [2].

Negative impacts of barren pens can be reduced by providing enrichment objects to the pigs, on which they can express some species-specific behaviors [3,7]. Other well-studied solutions include the provision of straw or other similar rootable materials, which have been shown to reduce aggression and enhance welfare (e.g., Moinard et al. [8]). However, the slurry system of fully or partly slatted pens is at risk of being blocked by straw or similar substrates [3,9]. Therefore, it is necessary to find alternative solutions such as enrichment objects. However, pigs' interest in the objects is likely to be influenced by the type of material and positioning within the pen [10,11]. Chewable and destructible materials, such as wood and rubber, are more likely to sustain interest compared to harder materials (e.g., metal chains) [10]. On the other hand, objects that are suspended at shoulder level to the pigs in the pen could initiate more manipulation since they are more visible and they stay cleaner than objects on the floor [7]. Enrichment objects associated with high levels of interest are more likely to reduce injurious social behaviors between pigs [3,11]. Therefore, tail- and ear-biting may be reduced by finding the most appropriate objects or combination of objects (e.g., Telkänranta et al. [12]). However, interest in an object can decrease after a few days [13,14] or a few weeks [15]. To prevent a rapid loss of interest, it is important to know the frequency at which objects should be replaced.

Additionally, soiled objects (by feces or dirt) potentially decrease pigs' interest [16] compared to clean objects, reducing their enrichment value. In a similar way, some highly destructible objects, such as ropes, can lose some of their attractiveness after a while because they are rapidly destroyed [12,17]. Although destructibility is seen as a beneficial characteristic for an enrichment object [10], it increases the frequency of object replacement. To be used on farms, enrichment objects should be practical and improve the economics of production [3]. Therefore, it is important to know whether washing or renewing objects is necessary to maintain their enrichment value.

This study was divided into two trials aimed at gathering information about the attractiveness of multiple enrichment objects presented to growing pigs. The objective of the first trial was to evaluate the short-term attractiveness of seven different objects based solely on their characteristics. The hypothesis was that destructible and chewable objects would trigger more manipulations. The objective of the second trial was to evaluate the effect of an object being soiled or damaged over time on its attractiveness. The hypothesis was that a clean or new object would be more attractive.

2. Materials and Methods

2.1. Animals and Housing

Large White × Landrace (or reciprocal mating) barrows were provided by one of the seven members of PigGen Canada from healthy multiplier farms and were group-housed in fully slatted pens measuring 2.6 m × 4.9 m in a research facility in Deschambault, Canada. Lights were on from 08:00 to 16:00 and no outside light could reach the pens. For the first trial, the average number of pigs per pen was 12 ± 3 (± SD), weighing 61 ± 9 kg on average, at the midpoint of the five-day recording periods. In total, 28 pens were used, representing 328 pigs. Due to a limited number of available pens containing pigs within the required weight range, eight pens were used twice, but never in a row. The weight range was 30 kg for trial one, meaning that all pens' average weight had to be between 45 kg and 75 kg during observations. The first trial lasted from May to August. For trial two, the average number of pigs per pen was 13.5 ± 1, weighing 47 ± 7 kg on average, at the midpoint of the five-day recording periods. Eight different pens and their pigs were also used twice because of a limited availability of pigs, but not for two consecutive observation periods, representing 107 different pigs. The weight range was 20 kg, meaning that all pens' average weight had to be between 37 kg to 57 kg during observations. The second trial lasted from November to December. For both trials, animals were fed a pelleted (corn- and soya-based) commercial diet and had ad libitum access to food and water. Two suspended metal chains were available in each pen before the beginning of the experiments and were removed during the trials. All animals were handled and cared for according to the guidelines of the Canadian Council on Animal Care [18], and the animal care committee at Laval University approved the procedures (CPAUL, protocol number 2015091).

2.2. Enrichment Objects

All objects (shown in Figure 1) were selected to cover many of the material characteristics (e.g., deformable, destructible, chewable, odorous) and locations in the pen (e.g., suspended, free in the pen, fixed on the floor) that have been reported to impact their degree of attractiveness [13]. Some objects were commercially available, whereas others were custom-made, as indicated in the list below:

- Ball: A commercial ball (Boomer Ball, Grayslake, IL, USA) made of rigid plastic and measuring 25 cm in diameter. The ball was loose in the pen and the pigs could play with it freely.
- Bite-Rite: A commercial object (Ikadan System A/S, Ikast, Denmark) suspended from the ceiling and comprising four chew sticks made of rubber-like material (2.2 cm in diameter, 25 cm in length) fixed to a central plastic cone. The chew sticks were at shoulder level to the pigs.
- Disc: A custom-made (Dundalk Plastic-Fab, Horning's Mills, ON, Canada) plastic disc (30 cm in diameter) suspended from the ceiling at shoulder level to the pigs. It had three hanging chains (15 cm in length) and three plastic strips (30 cm × 3 cm) fixed around it.
- Porcichew: A commercial object (Ketchum Manufacturing Ltd., Brockville, ON, Canada) made of a chewable plastic ring (15 cm in diameter) suspended from the ceiling at shoulder level to the pigs. The ring had a green apple scent.
- Rooting Cones: A commercial object (WEDA Dammann and Westerkamp GmbH, Goldenstedt, Germany) consisting of three chewable polyurethane balls (2 × 8 cm and 1 × 6 cm in diameter) fixed on top of metal springs (7 cm in length), which were mounted on a plastic ground plate anchored to the floor of the pen.
- Rope: A 30-cm-long polypropylene rope (Everbilt, Atlanta, GA, USA) suspended from the ceiling with a knot at the free end, which was at shoulder level to the pigs.
- Seesaw: A custom-made object (Agriculture and Agri-Food Canada, Sherbrooke Research and Development Centre, Sherbrooke, QC, Canada) consisting of two metal tubes forming a "T" (1.5 m in height, 1.2 m in width), which was fixed to the pen floor. Two polypropylene ropes (30 cm in length, 1.3 cm in diameter) were attached at both ends of a chain, which could slide inside the tilted tubes. The chain could slide for about 30 cm when a pig was pulling it from one side or the other. The tilted tubes could pivot 360° around the central axis. The ropes were presented at shoulder height of the pigs and had a knot at the free end to slow their destruction.
- Wood: A beam of untreated red cedar (10 cm × 10 cm × 30 cm) with a plastic ring (Jupiter Agro-Biotech, Saint-Hyacinthe, QC, Canada) on one end (4 cm in height, 25 cm in diameter) to elevate it from the floor for easier manipulations by the pigs and to reduce soiling. It was attached to a 1.5-m-long chain to limit its movements because the pigs could carry it to the feeders and block them.

Figure 1. Enrichment objects presented to growing pigs for five days in either one or both trials: (**a**) Ball; (**b**) Bite-Rite; (**c**) Disc; (**d**) Porcichew; (**e**) Rooting Cones; (**f**) Rope; (**g**) Seesaw; (**h**) Wood.

2.3. Experimental Procedures

The first trial was conducted in a fully randomized experimental design, repeating every treatment four times. Seven treatments were used, representing the seven enrichment objects (Ball, Bite-Rite, Disc, Porcichew, Rooting Cones, Seesaw, and Wood, as shown in Figure 1). Each week, four different objects were individually placed in four pens for five consecutive days. The objects were presented the afternoon before the recording started (day 0) to make sure that all the pigs were accustomed to the objects the next morning. Behavioral recordings were done on days 1 to 5 and on day 6, the objects were removed and cleaned. A group of pigs could not be used two weeks in a row for all eight of the pens that were used twice. Objects on the floor were placed in the cleanest part of the pen, far enough from the walls to allow all-round access. Suspended objects were also presented where the pigs could manipulate them from all sides. These locations were kept as similar as possible between weeks and pens.

The second trial was conducted in a complete block design and the effect of cleaning or replacing four selected objects was evaluated through a preference test, which was repeated four times. Two identical objects were presented at the same time for five consecutive days in one of four barren pens. One randomly selected object was not cleaned or replaced for five days (control object), whereas its duplicate was cleaned or replaced every morning from day 1 to 5 (treatment object). Treatment and control objects remained the same during those five days. As in trial one, the locations of every object were kept as similar as possible between repetitions.

To evaluate the effect of cleanliness, three commercial objects (Ball, Bite-Rite and Rooting Cones) were selected based on their characteristics and on information gathered in the first trial. The Ball and the Rooting Cones were shown to be rapidly soiled with feces and dirt as they were manipulated on the floor. In contrast, the chewing sticks on the Bite-Rite were only moderately soiled. Cleaning was performed using water and paper towels to remove dirt and feces on the Ball, chewing sticks of the Bite-Rite and three polyurethane balls, springs and ground plate of the Rooting Cones. No soap was used to avoid affecting the smell of the objects. Afterwards, objects were dried with paper towels. To determine the effect of object replacement, a suspended rope (shown in Figure 1) was chosen because of its rapidly destructible nature [17]. The treatment Rope was replaced with a new and identical rope daily.

During the treatment procedure (replacement or cleaning), both objects (control and treated) were taken outside the pen and were put back in at the same time, except for the Rooting Cones, because they could not be removed easily from the pen. Cleaning of the Rooting Cones was done in the pen, and the control Rooting Cones were also manipulated for the same amount of time, simulating a cleaning process. These procedures were performed to control for the potential effects of manipulation by employees on the degree of attractiveness of experimental objects.

2.4. Behavioral Observations

Behaviors were video recorded for both trials with cameras (TRENDnet: model TV-IP310PI, Torrance, CA, USA) fixed to the ceiling, providing a complete view around the objects. Recording was performed from 08:30 to 16:30 (when lights were on in the barn). A preliminary trial showed that pigs did not manipulate the objects much during darkness periods. Behavioral observation was performed by video analysis and focused on behaviors called manipulations, which were defined as any object-directed behaviors performed intentionally, involving a contact with the snout, head, legs (push, hit, rub) or with the mouth (chew, bite, pull, shake). Two variables were gathered directly from video analyses: total time spent manipulating the objects, calculated in terms of duration (rounded to the nearest second), and the frequency of manipulation from day 1 to 5. A third variable was the mean manipulation length and was obtained from the division of the duration by the frequency. Two manipulations were considered distinct when separated by a minimum of four seconds with no object-directed behavior performed by any pig (similar to the five seconds proposed by Gifford et al. [19]). A manipulation could be performed by multiple pigs simultaneously and it did not matter if the initial pig in contact with the object was not the same as the last one.

Behavioral sampling occurred over four continuous time periods of one hour each, equally distributed throughout the light period (09:00–10:00, 11:00–12:00, 13:00–14:00, 15:00–16:00) for the first trial, and three continuous time periods of half an hour each (09:30–10:00, 11:30–12:00, 13:30–14:00) for the second trial. The number and length of the time periods were respectively reduced and shortened for the second trial based on the results of the first trial, suggesting that reliable data could be collected in this way.

2.5. Statistical Analysis

Data analysis was performed with the MIXED procedure of SAS 9.4 (SAS Institute, Cary, NC, USA), using repeated measures. Pens were set as experimental units and the variables analyzed were the mean duration, mean frequency and mean manipulation length per time period. For the first trial, data on the object, day of observation, and the interaction between day and object, were analyzed as fixed effects and the repetitions as random effects. Multiple comparisons between objects were performed using Tukey's adjustments of Student's *t*-test. All data were tested for normal distribution.

For the second trial, preferences were evaluated using delta values between the same two objects (differences in duration, frequency and mean manipulation length, separately), calculated for every pen on each day (1 to 5). A delta different from zero (Student's *t*-test on each day) was interpreted as a preference. Manipulations of objects by the pigs on day 1 and each of the other days (2 to 5) were compared using Dunnett's correction for all objects independently, with control and treated data combined. Data on mean manipulation length in trial one and on the manipulation duration for the Bite-Rite in trial two had to be subjected to a logarithmic and a square root transformation, respectively, to ensure a normal distribution before the repeated measures analyses. Data are presented as least square means ± SEM, unless otherwise stated.

3. Results

3.1. Trial One

The frequency of manipulations (Figure 2a) was affected by object ($p < 0.0001$) and day ($p < 0.001$), and the interaction between object and day tended to be significant ($p = 0.07$). The Rooting Cones were the most frequently used objects by the pigs (21.0 times per hour on average), and the frequency of object manipulation was significantly reduced over time for the Ball, the Bite-Rite and the Disc between day 1 and day 5 (Figure 2a). There was an object–day interaction ($p < 0.05$) for the time spent manipulating objects (Figure 2b). The manipulation durations were consistently highest with Wood throughout the recording periods, while the percentage of time manipulating the Ball and the Rooting Cones decreased between day 1 and day 5 and between day 2 and day 5 for the Seesaw. An object–day interaction ($p < 0.05$) was also observed for mean manipulation length. The highest mean manipulation length was observed with Wood; it remained high over the five observation days, did not decrease significantly between day 1 and day 5 and remained higher than all other objects every day. The manipulation length with the Seesaw was higher on day 1 compared to day 5 (Figure 2c). The same has been found for the Ball and The Rooting Cones although they had lower manipulation lengths from day 1 compared with the Seesaw.

3.2. Trial Two

The pigs did not show any clear preference for control or treated (cleaned or replaced) objects. The only significant differences found were for the frequency of Ball manipulation on day 2 and the mean manipulation length of the Rooting Cones on day 3 (Table 1). Both indicated a preference for the control object and neither of these preferences persisted over time. The average frequency of manipulation for control and treated objects pooled together throughout the five experimental days were 19.2 ± 3.3 (Rope), 18.9 ± 1.5 (Rooting Cones), 16.7 ± 0.7 (Ball) and 9.9 ± 1.7 (Bite-Rite) manipulations per time period (30 min). The average percentages of time spent manipulating control and treated objects pooled together throughout the five experimental days were 32.4 ± 3.4 (Rope), 25.5 ± 1.8 (Rooting Cones), 16.7 ± 0.7 (Ball) and 13.6 ± 2.6 (Bite-Rite). Based on the frequency and duration data, the mean manipulation lengths in seconds were 45.8 ± 13.6 (Rope), 24.5 ± 8.9 (Bite-Rite), 23.2 ± 8.1 (Rooting Cones) and 15.2 ± 1.9 (Ball).

The four objects differed in their duration, frequency and mean length of manipulation over time (Figure 3). Interest was maintained at different relative levels between objects. The decrease in interest over time was observed between day 3 and day 5 and mainly affected the Ball and the Rooting Cones (Figure 3). Since no preferences were detected between control and treated objects (minus the two exceptions stated before), the pigs' interest in control and treated objects was considered to remain similar over the five-day periods.

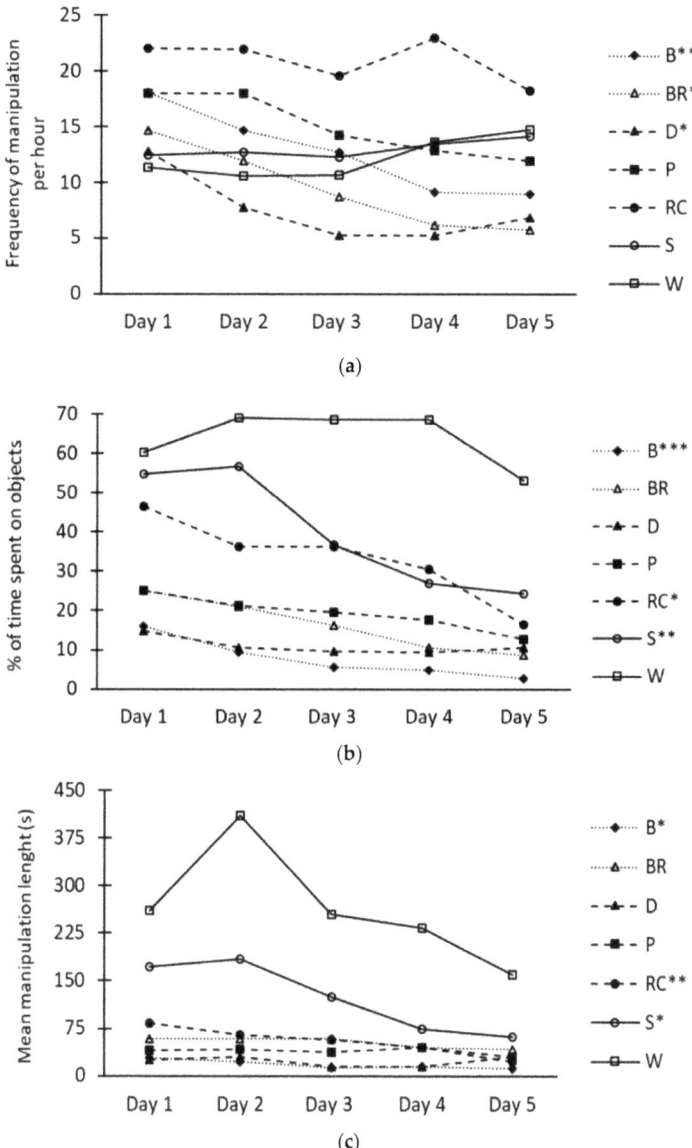

Figure 2. Daily adjusted means of the frequency of manipulation per hour (**a**), the proportion of time spent on objects (**b**) and the mean object manipulation length (**c**) in growing pigs. Objects (B: Ball, BR: Bite-Rite, D: Disc, P: Porcichew, RC: Rooting Cones, S: Seesaw and W: Wood) were presented individually in pens for five consecutive days. Significant effect of days is presented for each object separately with * ($p < 0.05$), ** ($p < 0.01$), *** ($p < 0.001$).

Table 1. Adjusted mean differences[1] between treated (cleaned Ball, Bite-Rite and Rooting Cones or replaced Rope) and control (untouched) objects for the frequency of manipulations (F), duration of manipulation in seconds (D) and mean manipulation length in seconds (ML) from day 1 to 5, when both objects were presented simultaneously to growing pigs.

Object	Ball			Bite-Rite			Rope			Rooting Cones		
	F	D	ML	F	D	ML	F	D	ML	F	D	ML
Day 1	−2.25	179.6	10.45	−3.00	−132.6	−4.61	8.63	13.0	−17.22	5.33	24.6	−6.92
Day 2	−5.50 *	19.6	4.85	0.33	−107.3	−8.59	3.08	−175.8	−26.99	1.17	57.2	4.94
Day 3	−1.75	103.1	6.42	−2.33	−285.1	−18.32	3.92	130.6	3.80	−2.58	−220.4	−6.18 **
Day 4	−0.83	−7.9	−0.63	−1.83	−290.1	−20.57	0.83	391.3	32.32	3.00	−99.2	−8.48
Day 5	−1.17	−27.3	−2.63	−2.67	−161.0	−1.02	4.75	96.8	−0.29	1.08	4.5	−0.34

[1] Negative differences (deltas) indicate more manipulations with control objects. * Indicates a significant preference for an object at $p < 0.05$, ** at $p < 0.01$ (pair-wise Student's t-test).

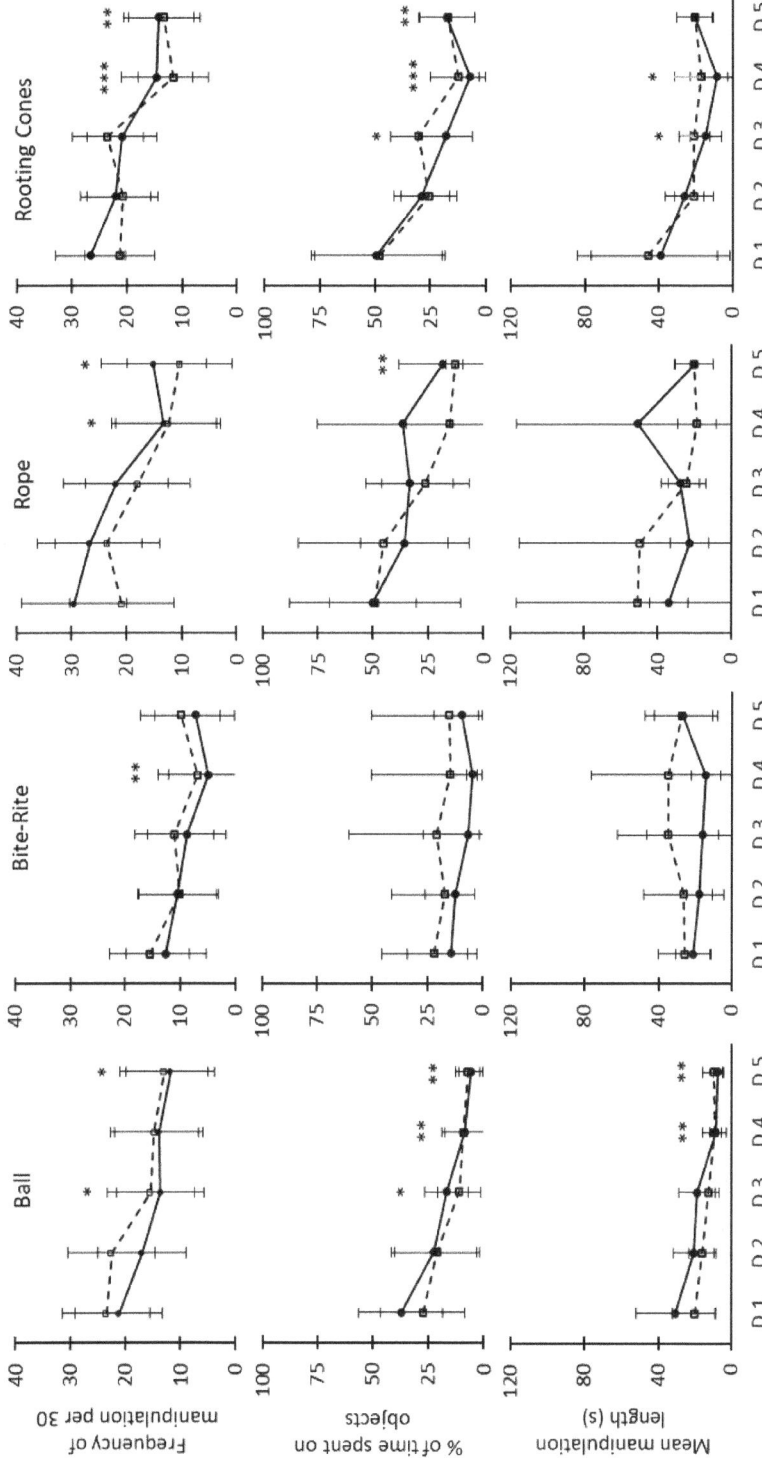

Figure 3. Adjusted means with 95% confidence intervals of frequency of manipulation per 30 min (upper row), percentage of time spent on manipulation (middle row) and mean manipulation length in seconds (lower row) for treated (solid line) and control (dash line) objects. Significant differences between daily average values (day 2 to day 5) from treated and control objects combined and the average value on day 1 (Dunnett's correction) is shown with * ($p < 0.05$), ** ($p < 0.01$), *** ($p < 0.001$).

4. Discussion

4.1. Behavioral Analysis—Trial One

One of the main findings of the present study was that more time was spent manipulating the Wood compared to any other object presented to the pigs. This difference was observed on every experimental day and the pigs' interest in the Wood did not decrease over the five-day period. Trickett et al. [17] have also reported such a sustained interest over time in dried wood. The wood beam has some of the key characteristics (e.g., destructible, chewable and odorous) known to stimulate pigs' interest [10]. Compared to the Rooting Cones, the Wood was manipulated less frequently, but for a longer time, giving it a longer mean manipulation length. Furthermore, the wood beams were presented directly on the floor and were not washed during experimentations. At the end of the five-day periods, they were soiled (personal observation), which is often associated with a decrease in interest [15,16]. However, the fact that the Wood was attached to a 1.5-m chain to limit its movements could have limited soiling and sustained the pigs' interest.

The Ball triggered the lowest manipulation duration of all objects. Although the manipulation frequency was relatively high, the mean manipulation length of the Ball was the shortest of all objects. This finding is in accordance with results discussed in a review by Bracke et al. [11], showing low levels of manipulation toward hard plastic balls. This can be explained by the incapacity of the pigs to chew or to simulate a rooting behavior on them. In addition, the Ball was the most soiled object after five days (personal observation), potentially decreasing its attractiveness [16]. The Rooting Cones were also soiled, but less than the Ball, since they could not be moved around the pen. Although immobile objects on the ground are sometimes seen as unfavorable to stimulate manipulations [15], it has been found that an object can be manipulated for a longer time and in a more vigorous way when presented in a fixed montage rather than on a free-moving montage (e.g., Courboulay [20], Elkmann and Hoy [21]). The fixed presentation of the Rooting Cones may have allowed the pigs to express their rooting and chewing behaviors more satisfyingly, hence the high frequency and time spent manipulating them (Figure 2a,b). The pigs could also manipulate the Rooting Cones while lying on the floor, giving them the opportunity to express natural behaviors when resting. The pigs could use the area where the fixed Rooting Cones were located as a resting area, increasing manipulation time and the cleanliness of the area and the objects themselves [20].

Using a meta-analysis, Averós et al. [7] showed that one of the characteristics that had a significant impact on how much time the pigs spent manipulating objects is if they are suspended. Objects presented that way are more visible and less susceptible to being soiled [15]. The suspended objects in the present study (Bite-Rite, Disc, Porcichew and Seesaw) resulted in different manipulation durations and frequencies. However, the mean manipulation lengths were similar between the Bite-Rite, the Disc and the Porcichew, although greater for the Seesaw. The greater length of time spent on an object for each individual manipulation observed from the Seesaw may be partly related to the destructive and deformable nature of its suspended ropes [10]. The knots were untied, and the ropes were mostly destructed within the five-day periods. Such a progressive destruction of an object could sustain pigs' interest in an object [22].

The Bite-Rite, Disc and Porcichew were subject to similar levels of manipulation. Low manipulation levels for the Bite-Rite were unexpected since the chewing sticks were deformable and destructible—two characteristics that seemed valued by pigs [10]. Chewing is one of the highly prioritized behaviors of pigs and a lack of chewable material is seen as a possible cause for harmful oral manipulations between pigs when placed in barren environments [12].

The Disc could have been less attractive to the pigs because it was hard to chew. Furthermore, the suspended chains around the disc could have been unattractive to the pigs as metal objects generally offer low enrichment benefits [11]. Chains have been associated with shorter manipulation times compared to other chewable objects, such as rubber hoses or ropes (e.g., Apple and Craig [23], Hill et al. [24]). It is also possible that the pigs were already accustomed to chains because all the pens had two suspended chains before experimentations began. The Porcichew was chewed more than the Disc, even though both objects were made of a similar hard plastic. The apple scent from the Porcichew could have helped to increase manipulations since odors are appealing to pigs [10].

In this research, analysis based on duration and mean manipulation length seemed to be more representative of the interest in the enrichment objects than the frequency of manipulation, according to the five-day manipulation patterns seen in Figure 1. This is in line with the proposition of Bracke [16] that an analysis solely based on parameters related to frequency could be insufficient to properly evaluate the welfare value of an enrichment object. For example, Telkänranta et al. [12] obtained similar manipulation frequencies for two enrichment objects, although only one of them reduced tail- and ear-biting in finishing pigs. The results from the present study offer different interpretations depending on how behavior was measured. Based on the frequency, the Rooting Cones would be the preferred enrichment object but based on time, the Wood would be due to both the duration and mean length of manipulation. However, the duration and mean length of manipulation with the Wood did not decrease over time, whereas these variables decreased significantly over time for the Rooting Cones, which suggests that the Wood may be a better option. Furthermore, the Ball and the Wood showed similar average frequencies of manipulation over five days, but the pigs spent over eight times as many seconds manipulating the Wood per manipulation. In addition, duration of manipulation changed differently over time between objects. The combination of both duration and frequency analyses, done as the mean manipulation length in this study, provides a more comprehensive explanation of the effects of enrichment objects on pigs' interest.

4.2. Behavioral Analysis—Trial Two

The main finding of this experiment is the absence of a clear effect of washing or replacing an object daily on the behavior of pigs over five days. The only preferences detected were an increased frequency of contact with the unwashed Ball on day 2 and a longer mean manipulation length on day 4 with the unwashed Rooting Cones. The preference for the unwashed Ball did not translate into an increased manipulation duration. This may be explained by the general low level of attraction for a hard-plastic ball [11], as was also observed in trial one. The more frequent manipulations with the control Ball and Rooting Cones could be explained by the curiosity of the pigs toward the smell of the dirty object. Odor is considered as a potential characteristic enhancing interest in the objects [10] but in this case, the smell of the dirty object could have stimulated the pigs to initiate small bouts of manipulation while they were smelling them, which does not necessarily translate into an indication of high interest (as discussed in the previous Section 4.1). It is also possible that the odor from handling by the people who were cleaning the objects could have triggered some manipulations on both treatment and control objects. However, no measurements were taken in the hour following the reintroduction of the object in the pen. Therefore, it is not possible to quantify this potential effect.

This trial showed that a daily wash or replacement of the selected objects was unnecessary in the short term, based on the preference test. However, it could be necessary to wash or replace some objects after a longer period. For example, the Rope would be destroyed to a point where it would be unusable after a long period (control Rope was not too heavily destroyed after five days but it is estimated that another 10–15 days of manipulations would have resulted in complete destruction). In the study by Trickett et al. [17], in which the pigs had continuous access to ropes, replacement was needed every two weeks (14 days) due to progressive destruction. In our study, it was noted that the Ball was completely covered in dirt and feces by the fifth day in general. Longer-term studies would be necessary to determine the potential negative effect of a dirty object on pigs' interest. Effects on pig health should also be measured, since dirty objects may represent a biosecurity risk [3].

The objects on the floor (Ball and Rooting Cones) were more affected by the decline of interest over time. These results are similar to those gathered in trial one and indicate that even when kept clean, those objects may not sustain pigs' interest sufficiently over time. The Bite-Rite also provided results similar to those of trial one. The degradation of the control Rope had no effect on pigs' interest, which as previously stated, could have been helpful to attract the pigs over time [22].

5. Conclusions

The level of pigs' interest in enrichment objects was dependent on the characteristics and the presentation of the objects. The interest seemed greater toward the destructible and chewable objects, but only the piece of wood sustained pigs' interest for the duration of the five-day period. Providing objects to growing pigs that have chewable and destructible properties could enhance their interest in them since they are capable of triggering species-specific behaviors and increasing the potential to reduce harmful contacts between pigs. However, those potential effects on pigs' welfare should be tested in a long-term trial. In addition to pigs' behavioral fulfillment, labor associated with the daily replacement or cleaning of objects can be minimized since it did not have a positive effect on their attractiveness. The characteristics of an enrichment object and the associated maintenance are especially important to optimize the impact of enrichment objects on welfare, while minimizing the associated labor cost.

Author Contributions: J.-M.B. wrote the initial draft and the final version of this paper. Project administration, funding acquisition, paper editing and overall project supervision were done by R.B., N.D. and J.-P.L. J.-M.B. collected data, supervised the experiment on farm and conducted data analysis.

Funding: This research was funded by Swine Innovation Porc within the Swine Cluster 2 (project 1237): Driving Results Through Innovation research program. Funding was provided by Agriculture and Agri-Food Canada through the AgriInnovation Program, provincial producer organizations and industry partners.

Acknowledgments: We would like to thank the employees at the Centre de développement du porc du Québec, namely Frédéric Fortin and Sébastien Turcotte for the planning, Louis Moffet and Hélène Mayrand who worked at the farm. Thanks are also addressed to the undergrad students (Alexandra Dumas, Gabrielle Dumas and Gabrielle Tardif) for their contribution to the project and to Steve Méthot, Marjolaine St-Louis from the Sherbrooke Research and Development Centre of Agriculture and Agri-Food Canada for the statistical analyses and technical help, respectively. The authors are grateful to John Harding, Mike Dyck, Jack Dekkers and Graham Plastow, along with PigGen Canada, for allowing the use of the animals from their research program on disease resilience to conduct this experiment.

Conflicts of Interest: The authors declare no conflict of interest. The funders had no role in the design of the study; in the collection, analyses, or interpretation of data; in the writing of the manuscript, or in the decision to publish the results.

References

1. Velarde, A.; Fàbrega, E.; Blanco-Penedo, I.; Dalmau, A. Animal welfare towards sustainability in pork meat production. *Meat Sci.* **2015**, *109*, 13–17. [CrossRef] [PubMed]
2. EFSA. Scientific report on the risks associated with tail biting in pigs and possible means to reduce the need for tail docking considering the different housing and husbandry systems. *Efsa J.* **2007**, *611*, 1–13.
3. Van de Weerd, H.A.; Day, J.E.L. A review of environmental enrichment for pigs housed in intensive housing systems. *Appl. Anim. Behav. Sci.* **2009**, *116*, 1–20. [CrossRef]
4. Broom, D.M. Welfare, stress, and the evolution of feelings. *Adv. Stud. Behav.* **1998**, *27*, 371–403.
5. Taylor, N.R.; Main, D.C.; Mendl, M.; Edwards, S.A. Tail-biting: A new perspective. *Vet. J.* **2010**, *186*, 137–147. [CrossRef] [PubMed]
6. Sinisalo, A.; Niemi, J.K.; Heinonen, M.; Valros, A. Tail biting and production performance in fattening pigs. *Livestig. Sci.* **2012**, *143*, 220–225. [CrossRef]
7. Averós, X.; Brossard, L.; Dourmad, J.-Y.; de Greef, K.H.; Edge, H.L.; Edwards, S.A.; Meunier-Salaün, M.-C. A meta-analysis of the combined effect of housing and environmental enrichment characteristics on the behaviour and performance of pigs. *Appl. Anim. Behav. Sci.* **2010**, *127*, 73–85. [CrossRef]
8. Moinard, C.; Mendl, M.; Nicol, C.J.; Green, L.E. A case control study of on-farm risk factors for tail biting in pigs. *Appl. Anim. Behav. Sci.* **2003**, *81*, 333–355. [CrossRef]
9. Day, J.E.L.; van de Weerd, H.A.; Edwards, S.A. The effect of varying lengths of straw bedding on the behaviour of growing pigs. *Appl. Anim. Behav. Sci.* **2008**, *109*, 249–260. [CrossRef]

10. Van de Weerd, H.A.; Docking, C.M.; Day, J.E.L.; Avery, P.J.; Edwards, S.A. A systematic approach towards developing environmental enrichment for pigs. *Appl. Anim. Behav. Sci.* **2003**, *84*, 101–118. [CrossRef]
11. Bracke, M.B.M.; Zonderland, J.J.; Lenskens, P.; Schouten, W.G.P.; Vermeer, H.; Spoolder, H.A.M.; Hendriks, H.J.M.; Hopster, H. Formalised review of environmental enrichment for pigs in relation to political decision making. *Appl. Anim. Behav. Sci.* **2006**, *98*, 165–182. [CrossRef]
12. Telkänranta, H.; Bracke, M.B.M.; Valros, A. Fresh wood reduces tail and ear biting and increases exploratory behaviour in finishing pigs. *Appl. Anim. Behav. Sci.* **2014**, *161*, 51–59. [CrossRef]
13. Zonderland, J.J.; Vermeer, H.M.; Vereijken, P.F.G.; Spoolder, H.A.M. Measuring a pig's preference for suspended toys by using an automated recording technique. *CIGR Ej.* **2003**, *V*, 1–11.
14. Guy, J.H.; Meads, Z.A.; Shiel, R.S.; Edwards, S.A. The effect of combining different environmental enrichment materials on enrichment use by growing pigs. *Appl. Anim. Behav. Sci.* **2013**, *144*, 102–107. [CrossRef]
15. Blackshaw, J.; Thomas, F.; Lee, J. The effect of a fixed or free toy on the growth rate and aggressive behaviour of weaned pigs and the influence of hierarchy on initial investigation of the toys. *Appl. Anim. Behav. Sci.* **1997**, *53*, 203–212. [CrossRef]
16. Bracke, M.B.M. Multifactorial testing of enrichment criteria: Pigs 'demand' hygiene and destructibility more than sound. *Appl. Anim. Behav. Sci.* **2007**, *107*, 218–232. [CrossRef]
17. Trickett, S.L.; Guy, J.H.; Edwards, S.A. The role of novelty in environmental enrichment for the weaned pig. *Appl. Anim. Behav. Sci.* **2009**, *116*, 45–51. [CrossRef]
18. CCAC. *CCAC Guidelines on: The Care and Use of Farm Animals in Research, Teaching and Testing*; Canadian Council on Animal Care: Ottawa, ON, Canada, 2009; Available online: https://www.ccac.ca/Documents/Standards/Guidelines/Farm_Animals.pdf (accessed on 15 January 2019).
19. Gifford, A.K.; Cloutier, S.; Newberry, R.C. Objects as enrichment: Effects of object exposure time and delay interval on object recognition memory of the domestic pig. *Appl. Anim. Behav. Sci.* **2007**, *107*, 206–217. [CrossRef]
20. Courboulay, V. Comment l'apport d'objets manipulables en hauteur et au sol influence-t-il l'activité des porcs charcutiers logés sur caillebotis intégral. *J. Rech. Porc.* **2004**, *36*, 389–394. (In French)
21. Elkmann, A.; Hoy, S. Frequency of occupation with different simultaneously offered devices by fattening pigs kept in pens with or without straw. *Livest. Sci.* **2009**, *124*, 330–334. [CrossRef]
22. Feddes, J.; Fraser, D. Non-nutritive chewing by pigs: Implications for tail-biting and behavioral enrichment. *Trans. Am. Soc. Agric. Eng.* **1994**, *37*, 947–950. [CrossRef]
23. Apple, J.K.; Craig, J.V. The influence of pen size on toy preference of growing pigs. *Appl. Anim. Behav. Sci.* **1992**, *35*, 149–155. [CrossRef]
24. Hill, J.D.; McGlone, J.J.; Fullwood, S.D.; Miller, M.F. Environmental enrichment influences on pig behavior, performance and meat quality. *Appl. Anim. Behav. Sci.* **1998**, *57*, 51–68. [CrossRef]

© 2019 by the authors. Licensee MDPI, Basel, Switzerland. This article is an open access article distributed under the terms and conditions of the Creative Commons Attribution (CC BY) license (http://creativecommons.org/licenses/by/4.0/).

Article

Rearing Undocked Pigs on Fully Slatted Floors Using Multiple Types and Variations of Enrichment

Jen-Yun Chou [1,2,3,*], Constance M. V. Drique [4], Dale A. Sandercock [2], Rick B. D'Eath [2] and Keelin O'Driscoll [1]

1. Pig Development Department, Teagasc, P61 P302 Moorepark, Ireland; keelin.odriscoll@teagasc.ie
2. Animal & Veterinary Sciences Research Group, SRUC, Roslin Institute Building, Easter Bush, Midlothian EH25 9RG, UK; dale.sandercock@sruc.ac.uk (D.A.S.); rick.death@sruc.ac.uk (R.B.D.)
3. Royal (Dick) School of Veterinary Studies, University of Edinburgh, Easter Bush, Midlothian EH25 9RG, UK
4. Agrocampus Ouest, 35042 Rennes, France; constance.drique@agrocampus-ouest.fr
* Correspondence: jenyun.chou@ed.ac.uk; Tel.: +353-83-488-4408

Received: 28 February 2019; Accepted: 28 March 2019; Published: 2 April 2019

Simple Summary: Floors with a series of gaps to allow pig manure to pass through (slatted floors) are common in the pig industry as they enable the efficient management of pig waste. Pigs need enrichment materials to occupy them and to reduce harmful behaviours like tail biting. Loose materials are problematic in slatted systems because they can block up the slats and slurry pumps. This study aimed to establish if it is possible to rear pigs with undocked tails in a fully slatted system with compatible enrichment, while keeping tail biting at a manageable level and to investigate how important the variation of enrichment is. The results showed that although some tail biting occurred, the level was low with only mild lesions observed. Only 1 out of 96 pigs sustained severe tail damage (portion of tail bitten off). Pigs receiving a more varied enrichment tended to have lower tail lesion scores than pigs continuously presented with the same enrichment. This study showed that an optimal quantity and quality of slat-compatible enrichment provision can reduce the risk of tail biting in undocked pigs in fully slatted systems. The important role of adequate environmental enrichment provision in fulfilling pigs' biological needs cannot be overemphasised.

Abstract: In fully slatted systems, tail biting is difficult to manage when pigs' tails are not docked because loose enrichment material can obstruct slurry systems. This pilot study sought to determine: a) whether intact-tailed pigs can be reared with a manageable level of tail biting by using multiple slat-compatible enrichment; b) whether a variation of enrichment has an effect; and c) whether pigs show a preference in enrichment use. Ninety-six undocked pigs were given the same enrichment items from one week after birth until weaning. At weaning, four different combinations of 8 enrichment items were utilized based on predefined characteristics. These were randomly assigned to 8 pens (n = 12 pigs/pen). Four pens had the same combination (SAME) from assignment and four pens switched combinations every two weeks (SWITCH). Individual lesion scores, interactions with the enrichment, and harmful behaviours were recorded. The average tail score during the experiment was low (0.93 ± 0.02). Only one pig in a SAME pen had a severely bitten tail (partly amputated). The overall level of interaction with enrichment did not decline over time. Pigs interacted with a rack of loose material most frequently ($p < 0.001$). The study showed promising results for rearing undocked pigs on fully slatted floors using slat-compatible enrichment.

Keywords: pig; environmental enrichment; slatted system; tail biting

1. Introduction

Tail biting is a damaging behaviour performed by pigs which causes injuries and pain in the recipients, with the worst cases causing permanent spinal infection and death [1]. The economic loss resulting from premature death and carcass condemnation due to tail biting is a serious issue for the pig industry [2,3]. Therefore, in order to control tail biting, tail docking is often practiced; there is a lower tail biting prevalence in docked than in undocked pigs [4]. However, routine tail docking is not permitted in the EU [5], and it can only be performed if other measures have been shown to be unsuccessful in controlling tail biting. It is, therefore, important to both encourage changes in attitudes and practices in the pig industry to ensure tail docking will be phased out and to provide practical and feasible advice to farmers on how to manage tail biting in different production systems.

Tail biting is most commonly observed when pigs are housed in suboptimal conditions (e.g., a high stocking density [6], high levels of competition for feed or water, and poor ventilation [7]). A lack of environmental stimulation for pigs has also been recognised as a key risk factor for tail biting [7]. In countries where tail docking is completely banned, loose materials are usually provided on the solid floor in fully concrete or part-slatted pens [8]. Fully slatted floors are, thus, considered a major risk factor for tail biting damage [9,10], at least in part because substrates are not usually placed on the floor as it could obstruct slurry storage systems [11]. Different organic enrichment materials that are not presented as litter and, therefore, more compatible with slatted systems have been investigated, mainly with docked pigs [12–15], but there has been limited evidence of efficacy where tail biting risk is high. Haigh et al. found that compressed straw blocks were no more effective than hanging rubber toys at reducing levels of damaging behaviour or tail lesion scores in docked pigs [14]. Likewise, other studies found no benefits to wood, when compared with a rubber floor toy [16]. Holling et al. also reported inconclusive effects on tail biting when using a straw dispenser due to its low occurrence [15], and Bulens et al. did not record any tail lesion when comparing different straw dispensers in docked pigs [12]. In undocked pigs, Veit et al. provided pigs with loose material (either corn silage or alfalfa hay) from the second week of life until 40 days post weaning and found that, although corn silage sustained pigs' interest and reduced tail lesion prevalence, the percentage of pigs with some tail loss was over 50% in all treatments by the end of the experiment [17]. Zonderland et al. showed that a small amount of long straw in a rack could reduce the presence of tail wounds although still not as effective as an even smaller amount of loose straw presented on the floor [18]. There was also an issue with the blockage of the manure system when using long straw in the straw rack.

The enrichment materials which pigs prefer to use are characterised by being destructible, odorous, chewable, deformable, and ingestible [19,20]. A common perception exists that enrichment devices or materials will be less preferred by pigs once they are no long novel and that, therefore, the frequency of use decreases [19–22]. Trickett et al. found that the rope always generated less interaction when it was renewed than when it was first introduced [22]. However, in most publications, the concept of "renew" and "novel" are often mixed and difficult to be distinguished between.

This study had three aims. The first was to investigate whether pigs with undocked tails could be reared from birth to slaughter without tail biting when multiple enrichment items compatible with fully slatted floors were provided. Enrichments were selected based upon a variety of predetermined characteristics and types of presentation. Our second aim was to investigate the importance of novelty in enrichment provision. Pigs were divided into two treatment groups postweaning: SAME and SWITCH. SAME pigs had the same age-appropriate enrichment for the weaner and finisher stages, while the SWITCH pigs had enrichment materials changed every two weeks. The number of weeks each enrichment combination was provided was kept equal between the SAME and SWITCH treatments overall. Finally, the third aim was to determine whether pigs showed any preference for the different materials provided.

2. Materials and Methods

2.1. Ethical Statement

Ethical approval was obtained from the Teagasc Animal Ethics Committee (TAEC163–2017). All procedures were carried out in accordance with the Irish legislation (SI no. 543/2012) and the EU Directive 2010/63/EU for animal experiments.

The primary ethical concern for this study was that there would be a high risk of tail biting outbreaks due to not docking the pigs' tails. To this end, pigs' tails were inspected twice a day during the week and once daily during the weekend by one of the first two authors of the study (J.-Y. Chou or C. Drique), and routine checks were also performed by the technical staff in the research centre. A tail biting outbreak was defined as when 3 or more pigs in a pen were observed to have fresh, clearly visible blood present on their tails and/or when 3 or more pigs in a pen had tail damage reaching score 3 on the tail scoring system over a consecutive 72 h. When an outbreak occurred, a set of predefined and randomised intervention protocols were deployed: 1. Removing victims, 2. removing suspected biters, or 3. adding in 3 ropes as additional enrichment and applying a layer of ointment (Cheno Unction, PharVet, Ireland) on all tails to reduce the smell of blood with a necessary treatment for the tail biting victims in place. In the case of pens with tail biting without meeting the outbreak criteria, the victims of severe biting were removed temporarily to a hospital pen for treatment for ethical reasons and reintroduced within a week. The hospital pens were located in the same room as experimental pens and had the same dimensions. As much as possible, pigs were accommodated with a partner from the same group to ease reintroduction after healing.

2.2. Animals and Housing

The experiment was carried out in the Pig Research Facility in Teagasc, Moorepark, in Ireland. Eight litters (Landrace × Large White) were used in the experiment, and all litters in the farrowing house were enriched with multiple enrichment materials. The piglets stayed in the farrowing pen (1.84 m × 2.50 m) for 4 weeks. One week after farrowing, a piece of hessian cloth (0.20 m × 0.20 m) and a small paper cup was provided, tied to the rear wall of the pen. A chewable plastic toy was put in at the beginning of the third week, and a piece of chopped bamboo (approximately 0.3 m in length, 0.05 m in diameter) at the beginning of the fourth week. The piglets were teeth-clipped, but their tails were kept undocked. No castration of the male piglets was performed. At weaning, the piglets were individually tagged and weighed, and 96 piglets were selected and randomly allocated to 8 treatment pens (12 pigs/pen) balanced for weight, sex (half male and half female), and litter mates (minimum 3 litter mates per pen). The mean weight between the pens and treatments was standardised as much as possible (minimum pen weight at 90.65 kg and maximum at 95.6 kg). The pigs were fed ad libitum with a standard pelleted diet through a single-spaced wet-dry feeding system, and water was also provided ad libitum via a nipple drinker. All pens were fully slatted with weaner pens dimensioned 2.4 m × 2.6 m and finisher pens 4.0 m × 2.4 m. The lighting was kept at 12 h per day to ensure a normal circadian rhythm (around 150 lux for weaners and 130 lux for finishers). The temperature was thermostatically maintained at 27 °C immediately postweaning and dropped 2 degree every 2 weeks thereafter, controlled by automatic heating and automatically controlled mechanical ventilation. The pigs stayed in the weaner house for 7 weeks before being transferred to the finishing house without remixing. The finisher house temperature was maintained at 20 °C with the same automatically controlled mechanical ventilation.

2.3. Enrichment Treatments

Eight enrichment categories were created based on the different properties of enrichment items suitable for pigs, as outlined by Van De Weerd et al. (2003) [19] (Table 1):

Table 1. The enrichment categories used in the experiment.

Categories	Properties [19]					
	Rootability	Durability	Edibility	Presentation	Texture	Location
1. Common item *	-	-	-	-	-	-
2. Floor toy	Yes	Deformable	Chewable	Movable	Soft	Floor
3. Wood post	Yes	Destructible	Edible	Attached	Hard	Floor
4. Hanging wood	No	Destructible	Edible	Suspended	Hard	Eye level
5. Loose material (long rack)	No	Renewed	Edible	Attached	Loose	Eye level
6. Fabric	No	Destructible	Chewable	Suspended	Soft	Eye level
7. Hanging chewtoy	No	Deformable	Chewable	Suspended	Soft	Eye level
8. Loose material (container)	No	Renewed	Edible	Suspended	Loose	Eye level

* The same item was present in all pens (an Easyfix® floor toy for weaners and a Piglyx® lick block for finishers).

Except the first category which was a common item present in all pens (an Easyfix® floor toy in the weaner stage and a Piglyx® lick block in the finisher stage; all items used in the experiment are listed in Table 2), four different enrichment items were identified, which fell into each category. All were easily and cheaply available in Ireland. The items within each category were then randomly assigned to one of four "combinations". Thus, in each "combination", there were 8 enrichment items (i.e., one from each of the 8 categories). To ensure the items could sustain wear and tear from use by larger pigs, more sturdy items were chosen for the finisher stage, but the characteristics of each category were consistent across stages.

Four pens of pigs were assigned to the "SAME" treatment. Each of these pens was assigned to a different one of the four "combinations" for the duration of the weaner and finisher stage. Four other pens of pigs were in the "SWITCH" treatment. For these pens, the "combination" they received was switched every two weeks. As such, the number of weeks each "combination" was provided was kept equal between the "SAME" and "SWITCH" treatments overall (see Appendix A for the detailed treatment assignment and distribution between pens and weeks).

During the experiment, the enrichment was monitored, cleaned, renewed, or replaced twice a day to ensure that pigs had access to all items at all times. The weight of the loose materials provided in a rack was recorded whenever it was renewed.

A previous study conducted on the same research farm by the authors showed that using a single enrichment item to rear undocked pigs under similar conditions led to a high occurrence of tail biting outbreaks with a large amount of pigs suffering from tail amputation [23]. Thus, due to ethical concerns and the legal requirement to provide pigs with manipulable materials, no negative control was used in the current experiment.

Table 2. The enrichment item combinations from different categories in randomised groups.

Stage	Weaner				Finisher			
Combinations/Categories	1	2	3	4	1	2	3	4
1. Common item	Easyfix® floor toy	Easyfix® floor toy	Easyfix® floor toy	Easyfix® floor toy	PigLyx®	PigLyx®	PigLyx®	PigLyx®
2. Floor toy	Brush	Dog toy	Rubber boot	Easyfix floor toy	Larch	Easyfix floor toy	Cherry	Spruce
3. Wood post	Larch (M)	Pine	Spruce	Larch (S)	Spruce	Pine	Larch (S)	Larch (M)
4. Hanging wood	Pine	Bamboo	Larch	Spruce	Pine	Larch	Spruce	Bamboo
5. Loose material (long rack)	Grass	Straw	Sawdust	Shredded paper	Grass	Straw	Miscanthus	Shredded paper
6. Fabric	Cotton piece	Cardboard roll	Coconut basket	Hessian cloth	Cardboard roll	Tonne bag	Hessian sack	Astroturf
7. Hanging chewtoy	Tennis balls	Rubber pipes	Dog tug toys	Bamboo pieces	Rubber boot	Dog toy frisbee	Bamboo stick	Dog tug toy
8. Loose material (container)	Sawdust powder	Shredded grass	Peat	Chopped carrots	Shredded paper (S)	Sawdust powder	Peat	Dried grass

2.4. Pig Assessment

2.4.1. Physical Measures

The pigs were weighed individually at weaning, upon moving to the finisher house, and before slaughter using weighing scales calibrated to the expected weight range of the pigs. Every 4 weeks, they were also weighed as a group to monitor the growth rate. The pen level feed intake was recorded through the computerized feeding system (BigFarmNet Manager, Big Dutchman Ltd. v3.1.5.51039, Vechta Calveslage, Germany).

Every two weeks, tail lesions, ear lesions, and tear staining levels were checked and scored on each pig individually by a single observer. The tail lesions were scored using the scoring system developed by the Farewelldock consortium [24]. One score was given according to the severity of the tail lesions observed (0: no lesion, 1: bite marks, 2: open wound, and 3: swollen bite wounds), and a second score was given based on the presence of blood (0: no blood, 1: black scar, 2: older red blood, and 3: fresh blood). Ear lesions (0: no lesion, 1: superficial scratches, 2: evidence of recent bleeding, 3: substantial cuts and bleeding, and 4: part of the ear missing) were scored with the same method detailed in Chou et al. [13]. Tear stains were recorded with the DeBoer–Marchant–Forde Scale (score 0–5), and both eyes were scored independently [25]. The pigs reached market age at 114 days postweaning. They were individually marked with slap marks the day prior to slaughter and traced through the factory to obtain postmortem data from the carcasses. The tails were scored on the slaughter line after scalding by a single trained observer, using a 0–4 scale modified from Harley et al. [26].

2.4.2. Behaviour

- Direct behaviour observations

Observations of animal behaviour took place on two different days per week (except immediately postweaning and in the week after transferring to the finisher house) and twice per day: once in the morning around 10:30 h and once in the afternoon around 14:30 h, using a predefined ethogram (Table 3). These times were chosen based on a previous study [23] which indicated that this was when the pigs were most active. The pigs' interactions with the enrichment items, their behaviours towards other pen mates, and play behaviour were the focus of the observations. Behaviour sampling was conducted continuously for 5 min in each session. All pigs were identified individually by a unique colour pattern marked on their back. The pigs were marked twice a week on the day before the observation day in order to minimise the disturbance.

- Video recording and behaviour scanning

All pens were video recorded (by QVIS HDAP400 CCTV cameras and Pioneer-16 digital recorder case, Hampshire, UK) continuously on one day every week, except for immediately postweaning and in the week after moving to the finisher house. Behavioural observations were carried out using scan sampling. A static image from the video recording was obtained every 20 min between 07:00 h and 20:00 h ($n = 39$ data points). The number of pigs interacting with each of the 8 enrichment items during that instant was counted. The identity or sex of the pigs was not recorded.

Table 3. The ethogram used for behaviour observation.

Enrichment Directed Behaviours	Description of Behaviours
Bite device	Oral manipulation of the device with mouth open
Root device	Manipulation of the device by manoeuvring the device locomotively using the snout
Aggressive encounter	Biting, headknocking, or pushing over access to the device
Other	Any physical contact with the device other than mouth/snout
Negative behaviours	
Tail manipulation not at feeder (standing or lying/sitting)	Oral manipulation of the tail of another pig away from the feeder.
Tail manipulation at feeder	Oral manipulation of the tail of another pig which is feeding in the vicinity of the feeder.
Ear manipulation (standing or lying/sitting)	Oral manipulation of the ear of another pig.
Biting other parts of the body	Biting pen mate in another region other than tail and ear, e.g., hock, flank, snout, or genital area
Belly nosing	Rubbing/manipulating a pen mate's belly/flank region with rhythmic up and down snout movement
Mounting	Putting two front legs on top of another pig
Aggressive behaviour	Pushing, headknocking, and open-mouth fighting with pen mates
Positive behaviours	
Social nosing (face)	Gentle, non-open mouth nosing on another pen mate's facial area (without aversive reaction from the recipient)
Individual play	Based on Newberry et al. (1988) [27] and Donaldson et al. (2002) [28], any scampering, pivoting, head tossing, flopping, and pawing movement.

2.5. Statistical Analysis

The data were analysed using Statistical Analyses System (SAS, version 9.4, 1989, SAS Institute Inc., Cary, NC, USA). All residuals were checked for normality after analysis, and an appropriate transformation was applied (square root transformation) to non-normally distributed frequency data. For continuous and normal data such as the weight of the pigs, feed intake, the weight of loose materials, and behavioural frequency data, the MIXED procedure was used to fit mixed models. The stage (weaner/finisher), week within stage, treatment (SAME/SWITCH), and the enrichment combination within each stage were included as fixed effects in all behaviour analyses. When analysing the level of interaction with different enrichment items, category and the interaction between category and treatment were also included as additional fixed effects to investigate how often pigs interacted with each category. Direct behavioural observations were analysed as frequencies per pig per minute, and the interactions from the video scans were analysed as frequencies per pig per scan. For the analysis of the consumption of loose materials, week, treatment and the interaction between treatment and material were included. For the direct observations, sex was included as a fixed effect to investigate differences between male and female pigs. The behaviour data were averaged with week, which was included in the model as a repeated effect. All behavioural data were analysed using pen as the experimental unit ($n = 8$), except when comparing sex differences ($n = 16$). The growth data such as weight and feed intake were also analysed at the pen level ($n = 8$), as well as the weight of the loose materials ($n = 8$).

All physical scores (alive and postmortem) were analysed using the GLIMMIX procedure to fit the Generalised Linear Mixed Models, including week (or stage), treatment, and sex as fixed effects. This was analysed using individual pig as the experimental unit ($n = 96$).

3. Results

3.1. Consumption of Materials

Pigs in the SAME treatment consumed more grass than those in the SWITCH treatment in the weaner stage (SAME 104.9 ± 5.7 vs. SWITCH 28.6 ± 8.5 g/day, mean ± S.E., $p < 0.001$). Likewise, they consumed more grass (SAME 170.7 ± 7.6 vs. SWITCH 63.2 ± 10.8 g/day, mean ± S.E., $p < 0.001$) and miscanthus (SAME 234.7 ± 7.3 vs. SWITCH 130.4 ± 7.8 g/day, mean ± S.E., $p < 0.001$) in the finisher stage than the SWITCH pigs.

The replacement rate of other items only included the SAME groups since the items were always replaced in the SWITCH groups every 2 weeks according to the experimental design. Descriptive results on the frequency of replacement of the fabric materials (category 6) and the materials provided in the container (category 8) are provided in Table 4.

Table 4. The descriptive replacement rate for the fabric and loose material provided in the container in two different stages among the SAME groups.

Stage		Weaner				Finisher		
Category/Group	Fabric	Days (mean ± S.E.)	Container	Days (mean ± S.E.)	Fabric	Days (mean ± S.E.)	Container	Days (mean ± S.E.)
1	Cotton	16.3 ± 14.3	Sawdust	2.2 ± 0.4	Cardboard	2.4 ± 0.5	Paper	62.0 *
2	Cardboard	4.9 ± 1.8	Grass	2.6 ± 0.4	Tonnebag	64.0 *	Sawdust	8.9 ± 4.3
3	Coconut	5.4 ± 1.3	Peat	1.2 ± 0.1	Hessian	10.7 ± 3.6	Peat	8.9 ± 3.4
4	Hessian	9.8 ± 3.3	Carrot	1.8 ± 0.2	Astroturf	8.0 ± 4.1	Dried Grass	62.0 *

* The items were not replaced during the duration of the trial, and therefore, no S.E. is available.

3.2. Physical Measures

Out of the 96 pigs used in the experiment, only one pig (from a SAME pen) received a severe tail biting injury during week 4. This pig experienced a removal of 2/3 of the tail and was removed from the experiment.

Although in both treatments tail lesion scores were low, there was a nonsignificant tendency for them to be numerically higher in the SAME pigs (Table 5). There was no difference in ear lesion or tear staining scores between the SAME and SWITCH pigs or in the postmortem tail lesion scores (Table 5). Neither was there an effect of sex on antemortem lesion scores or postmortem tail lesion scores. In terms of growth, the pigs in both treatments had the same average daily gain (ADG) and feed conversion ratio.

Table 5. The lesion scores and tear stains for the SAME and SWITCH groups: The score data are presented as mean ± S.E. (n = 48 per treatment group).

Scores	SAME	SWITCH	F-Value	p-Value
Tail (alive)	0.83 ± 0.03	0.76 ± 0.03	3.43	0.07
Ear	0.98 ± 0.02	0.98 ± 0.02	0.01	>0.05
Tear staining	1.83 ± 0.07	1.79 ± 0.07	0.2	>0.05
Tail (postmortem)	1.00 ± 0.15	0.80 ± 0.13	1.01	>0.05

Female pigs had lower ADG than male (0.89 ± 0.01 vs. 0.97 ± 0.01 kg/day; $F = 20.74_{(85,1)}$, $p < 0.001$) and lower live weights recorded before slaughter (female 107.58 ± 1.34 vs. male 116.45 ± 1.41 kg; $F = 25.64_{(85,1)}$, $p < 0.001$).

The pigs had lower tail lesion scores in the weaner than in the finisher stage (0.72 ± 0.02 vs. 0.89 ± 0.02; mean ± S.E., $F = 41.31_{(658.1,1)}$, $p < 0.001$), whereas ear lesion scores were worse in the weaner stage than later (1.02 ± 0.02 vs. 0.95 ± 0.02; mean ± S.E., $F = 6.04_{(667.1,1)}$, $p < 0.05$). Figure 1 shows the pattern over time; tail lesion scores gradually increased from weaning until movement to the finishing stage, and then they stabilised ($p < 0.001$). Ear lesion scores were reasonably consistent across time, only lower in week 14 than in weeks 4 and 8 ($p < 0.01$, Figure 1).

3.3. Direct Behavioural Observations

There was no effect of the treatment (SAME or SWITCH) on any measure of behaviour.

3.3.1. Enrichment Directed Behaviour

Table 6 summarises the amount of interactions with the enrichment overall and within each category in both the weaner and finisher stages. The pigs in the weaner stage were observed to interact more with enrichment in general, driven by more interaction with the common item (Easyfix for weaners and PigLyx for finishers) and the wood post. The only item for which there was less interaction in the weaner than finisher stages was the loose material in the container. There was no effect of sex on the amount of interaction with the enrichment.

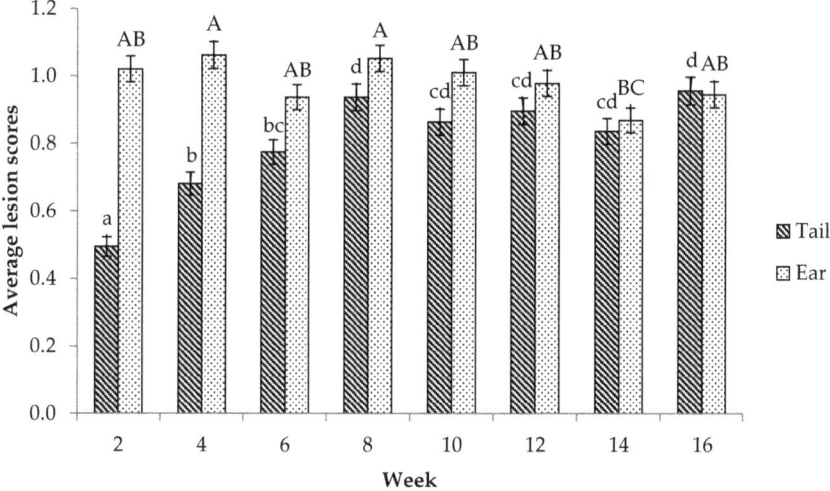

Figure 1. The mean ± S.E. tail and ear lesion scores across time (n = 48 per treatment group): The tail lesion score was the average of the 2 tail scores given, where one score represented the severity of the lesions (0: no lesion, 1: bite marks, 2: open wound, and 3: swollen bite wounds), and the other represented the presence of blood (0: no blood, 1: black scar, 2: older red blood, and 3: fresh blood). Ear lesions were scored on a 0–4 scale (0: no lesion, 1: superficial scratches, 2: evidence of recent bleeding, 3: substantial cuts and bleeding, and 4: part of the ear missing).

Table 6. The pig interaction with different enrichment categories during the weaner and finisher stages via direct observation and video scans: The data presented are frequencies per pig per minute at a pen level, as the mean ± S.E. (n = 8).

Categories	Direct				Video			
	Weaner	Finisher	F-Value	p-Value	Weaner	Finisher	F-Value	p-Value
All items	0.288 ± 0.012	0.253 ± 0.011	5.34	<0.05	0.083 ± 0.003	0.053 ± 0.002	75.52	<0.001
Common item	0.043 ± 0.003	0.019 ± 0.003	29.26	<0.001	0.014 ± 0.001	0.006 ± 0.001	44.98	<0.001
Floor toy	0.039 ± 0.006	0.035 ± 0.005	0.42	>0.05	0.013 ± 0.001	0.008 ± 0.001	10.77	<0.01
Wood post	0.016 ± 0.002	0.007 ± 0.002	9.17	<0.01	0.006 ± 0.001	0.002 ± 0.001	27.01	<0.001
Hanging wood	0.012 ± 0.002	0.009 ± 0.002	1.51	>0.05	0.002 ± 0.000	0.002 ± 0.000	0.01	>0.05
Loose material (long rack)	0.101 ± 0.011	0.096 ± 0.010	0.33	>0.05	0.024 ± 0.002	0.019 ± 0.002	3.98	<0.05
Fabric	0.039 ± 0.007	0.048 ± 0.006	1.35	>0.05	0.016 ± 0.002	0.009 ± 0.002	8.23	<0.01
Hanging chewtoy	0.023 ± 0.003	0.014 ± 0.002	5.63	<0.05	0.006 ± 0.001	0.003 ± 0.001	9.95	<0.01
Loose material (container)	0.015 ± 0.004	0.024 ± 0.003	6.68	<0.05	0.002 ± 0.001	0.005 ± 0.001	8.83	<0.01

3.3.2. Preferences between and within Each Category

During the entire experiment, the pigs interacted more with the material in the rack than any other item and the least with the wood post and hanging wood (Figure 2, $p < 0.001$). Within each combination, a preference for certain items was also observed. Within the "floor toy" combination, weaners interacted more with the brush than with the dog toy (0.07 ± 0.01 vs. 0.01 ± 0.01; mean ± S.E., $F = 3.82_{(83.6,6)}$, $p < 0.01$). In the finishing stage, the pigs were observed to interact most frequently with the cardboard roll among the fabric materials provided (cardboard = 0.07 ± 0.01, tonne bag = 0.04 ± 0.01, hessian sack = 0.04 ± 0.01, astroturf = 0.01 ± 0.01; mean ± S.E., $F = 5_{(56.7,6)}$, $p < 0.001$), and they also interacted more with the sawdust than the dried grass provided in the hanging container (0.04 ± 0.01 vs. 0.01 ± 0.01; mean ± S.E., $F = 2.71_{(77.9,6)}$, $p < 0.05$).

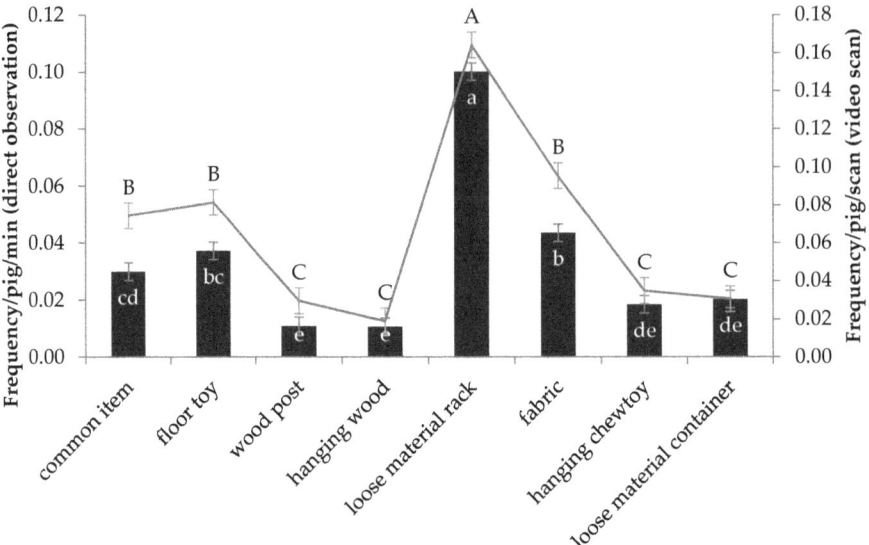

Figure 2. The overall pig interaction frequency with each category of enrichment at a pen level (n = 8) via direct observation (bar chart, different small letters denote significant difference) and video scans (line chart, different capital letters denote significant difference). The data are presented as the mean ± S.E.

3.3.3. General Pig Behaviour

A tail biting behaviour was performed more in the weaner (0.013 ± 0.002 instances/pig/min) than finisher (0.006 ± 0.001 instances/pig/min; $F = 16.2_{(177,1)}$, $p < 0.001$) stages and by females (0.011 ± 0.001 instances/pig/min) more often than males (0.007 ± 0.001 instances/pig/min; $F = 17.86_{(7,1)}$, $p < 0.01$). Likewise, ear biting was performed more in the weaner (0.025 ± 0.004 instances/pig/min) than finisher stages (0.020 ± 0.003 instances/pig/min; $F = 5.2_{(180,1)}$, $p < 0.05$). Regarding bites to the other parts of the body (e.g., snout, legs, etc.), there was no effect of stage, but females performed more of this behaviour than males (0.0139 ± 0.013 vs. 0.0084 ± 0.013 instances/pig/min; $F = 8.81_{(14,1)}$, $p = 0.01$). However, when all damaging behaviours were considered together (tail + ear + other body part directed), there was only a tendency for an effect of stage (weaner 0.060 ± 0.005, vs. finisher 0.047 ± 0.005 instances/pig/min; $p = 0.06$), and the effect of sex was just significant (Female = 0.063 ± 0.006, vs. male = 0.044 ± 0.006 instances/pig/min; $F = 4.48_{(14,1)}$, $p = 0.05$).

Similar to damaging behaviours, pigs performed more positive behaviours (sum of social nosing and play) in the weaner than finisher stages (0.031 ± 0.003 vs. 0.022 ± 0.002 instances/pig/min; $F = 10.09_{(180,1)}$, $p < 0.01$), and females performed more than males (0.029 ± 0.002 vs. 0.023 ± 0.002 instances/pig/min, respectively, $F = 7.93_{(14,1)}$, $p = 0.01$). Aggression was also higher in the weaner stages than finisher (0.103 ± 0.007 vs. 0.054 ± 0.005 instances/pig/min, respectively; $F = 50.42_{(180,1)}$, $p < 0.001$), yet contrary to all other behaviours, males performed more aggression than females (0.088 ± 0.006 vs. 0.069 ± 0.006 instances/pig/min; $F = 5_{(14,1)}$, $p < 0.05$).

3.4. Video Scans

Unlike the direct observations, the SAME pigs were recorded interacting with loose materials in the rack more frequently than the SWITCH pigs (0.20 ± 0.01 vs. 0.13 ± 0.01, mean ± S.E., $p < 0.001$) by video scans.

Similar to the results from the direct observations, the video scans also revealed that weaner pigs interacted more with the enrichment than finishers (Table 6). In this case, it was driven by an increased level of interaction with all items other than the hanging wood (no difference) and the loose material in a container, which, again, was interacted with less in the weaner than finisher stages. Over the entire experiment, similar to the direct observations, the rack of loose material received the most interactions, and the wood received the least (Figure 2; $p < 0.001$).

Between different types of floor toy, weaners interacted more with the brush than with the dog toy (0.020 ± 0.003 vs. 0.004 ± 0.003; mean $\pm$ S.E., $F = 4.01_{(53.7,6)}$, $p < 0.01$) and more with the Easyfix floor toy than the dog toy (0.018 ± 0.003 vs. 0.004 ± 0.003; mean $\pm$ S.E., $p < 0.05$). In the "hanging chew toy" category, pigs in the weaner stage interacted more with bamboo pieces than with tennis and the dog tug toy (bamboo = 0.010 ± 0.002, tennis = 0.003 ± 0.002, dog toy = 0.003 ± 0.002; mean $\pm$ S.E., $F = 3.33_{(85.6,6)}$, $p < 0.05$), and in the "container" category, they interacted more with peat than with the ground sawdust and grated carrot (peat = 0.0070 ± 0.0013, ground sawdust = 0.0003 ± 0.0013, dog toy = 0.0010 ± 0.0013; mean $\pm$ S.E., $F = 2.94_{(57.6,6)}$, $p < 0.05$).

4. Discussion

This study investigated the possibility of successfully rearing pigs without tail docking by using multiple slat-compatible enrichment items; we explored the effect of variation in enrichment provision and further compared pigs' preferences for different enrichment items. The results showed that, although a low level of tail biting occurred, the lesions were mild (lesion scores around 1 or less on a 0–3 scale) and that there were no tail biting outbreaks. Overall, providing pigs with a more varied enrichment had a marginally positive effect on the severity of tail lesions, whereby pigs given more variation tended to have lower tail lesion scores, although this was not statistically significant. Pigs also showed clear preferences for certain enrichment items, with the loose materials in a rack being most preferred.

4.1. Overall Effect of Enrichment On Tail Biting

Throughout the experiment, only one pig was severely tail bitten, with 2/3 of its tail amputated. No tail biting incidents reached the outbreak severity criteria set out beforehand. However, tail biting behaviours were still observed, and tail lesions scores became progressively worse over time, which is a characteristic of tail biting behaviour. The increase in tail lesion scores is more likely due to the low levels of biting on an ongoing basis, which were not acutely damaging to the recipients. The results demonstrated that it was possible to keep tail biting at a manageable level even among intact tail pigs on fully slatted floors by a high standard of enrichment provision and management that was compatible with the housing system.

4.2. Differences between SAME and SWITCH Groups

There was no difference in the general behaviour of the pigs that received a consistent type of enrichment (SAME) and pigs that had their enrichment types switched regularly (SWITCH) even though the SAME pigs had a slight but not statistically significant tendency to have worse tail lesions. Scott et al. suggested that it is the "stimulus properties" of the enrichment rather than the quantity alone that is important to sustain the interest of pigs [29]. In the current study, the SWITCH pigs could be more stimulated by the change in enrichment and, therefore, did not tail bite as much as the SAME pigs.

Although the level of interaction with the enrichment did not differ between the SAME and SWITCH pigs when observed directly, the results from the video observations showed that the SAME pigs interacted with the loose materials in the rack more than the SWITCH pigs. Moreover, when provided with grass, the SAME pigs consumed more than the SWITCH pigs. The loose materials were replenished whenever the rack was emptied, and a higher replenishment rate could have increased the novelty effect for the SAME pigs, which in turn stimulated more use of the rack. Van de Weerd et al.

found that the trait "renewed" was desirable for pigs and generated more interaction [19]. Trickett et al. also demonstrated that, whenever a point source enrichment item was replenished and reintroduced, it regenerated a higher frequency of interaction than a pre-existing item [22]. Thus, the attractiveness of the item and the effect of novelty might have an additive effect on the level of interaction. As the SWITCH pigs were exposed to multiple new enrichment items every two weeks, the novelty effect of all the items present in the pen may have diverted their attention away from the loose materials in the rack, even though, overall, they were still the most preferred by the pigs. The reason why the difference was only picked up by video scanning could be due to the observations including the whole day, rather than focused on the most active times of day.

4.3. Pigs' Preference Forenrichment Categories and Items

Regardless of age or treatment, pigs interacted with the loose materials in the rack most frequently; this is in agreement with previous research showing that loose straw or straw in racks occupies pigs' time more than other items [29–32]. Nevertheless, there have been some studies where pigs interacted with point-source enrichment (a straw pellet dispenser) more than loose materials (chopped straw/miscanthus) provided as floor litter [33]. Guy et al. also found that pigs preferred rope and chain to sawdust presented in a trough [34]. In the current experiment, there was no available litter on the floor. Thus, the material in the rack was likely the next best substitute and the most appealing for the pigs regardless of which type of materials used. The elevated rack we used kept the materials clean and easily accessible (up to 6 pigs could use it simultaneously).

Second to the loose materials in the rack, the floor toys and fabrics were the categories that pigs interacted with most. The floor toys were moveable, which increased the accessibility compared with the items that were installed at a fixed location. Scott et al. also mentioned social facilitation, whereby pigs were stimulated by pen mates, and therefore, synchronising their behaviour could increase pigs' interaction with enrichment [29]. The fabric, on the other hand, was soft and destructible, which were properties known to be favourable to pigs [19]. In contrast, both the wooden post and hanging wood attracted the least attention, which agreed with a previous study which compared wood blocks with ropes [22]. Telkänranta et al. found that pigs preferred fresh wooden pieces to chains but no difference between wood and polythene pipes [35]. The wood used in the current trial was not freshly cut but commercially dried, which could have reduced the pigs' interest.

The bamboo piece (compared with tennis ball, rubber pipe, and dog tug toy) and cardboard roll (compared with cotton piece, coconut basket, and hessian cloth), which were preferred by the pigs, both possess relatively more attractive characteristics compared to other items in their categories. Van de Weerd et al. identified the main characteristics of objects which captured and sustained the interest of pigs to be ingestible, odorous, chewable, deformable, and destructible [19]. A study also found that pigs interacted more with softer wood species than the hard [13]. In the current experiment, items that were more easily destroyed (destructible) also generated a higher amount of interaction and needed to be replaced more often. Another factor influencing the use of enrichment is cleanliness. Items that are easily soiled by excretion were used less by pigs [30]. Based on the experimenter's experience, the dog toy was constantly soiled and needed to be cleaned compared with the brush, rubber boot, and Easyfix® floor toy, which may have contributed to its low rate of use. It was smaller and lighter than the other items, so was more often covered in faeces when presented on the floor, which could have inhibited its use by the pigs.

The location of enrichment, particularly with regard to point source items which are fixed in a specific location in the pen, could affect the access and competition [30]. In the current study, the loose materials provided in the container were presented differently in the weaner and finisher stages: an open-top bucket hung at the side of the gate near the feeder for the weaners and a sealed canister suspended in the centre of the pen was used for the finishers. The finishers interacted with this category more than the weaners. This is consistent with the suggestion by Van de Weerd et al. [30] that, even when all other properties of the enrichment were the same, a central location could invite easier access and, therefore, more interaction than an item placed peripherally. Thus, factors such as location in the pen and presentation method need to be considered carefully when choosing environmental enrichment options for pigs.

4.4. Effects of Age and Sex

Weaners performed both more positive and harmful behaviours than finishers and interacted more with the enrichment, which could be due to pigs being more active in general when they are younger. Indeed, the video observations showed that weaners had more frequent interactions with most categories of enrichment items than finishers, except for the hanging wood. Docking et al. also found an effect of age of the pig on the level of interaction with objects, with finishers showing less interaction with objects than weaners [36]. Other studies have also shown that ear biting behaviours decreased as time progressed in the later weaner stage [37].

Nevertheless, tail lesion scores of finishers were marginally higher than those of weaners, which again agrees with previous studies [38]. This could be due to the accumulation of tail biting damage over time.

In the current study, female pigs performed more biting behaviour in general than males. As the females were lighter and had a lower average daily gain than males, they perhaps used biting behaviours to gain access to enrichment or the feeder [14]. Although some studies have shown that male pigs tended to have more tail lesions than females [39], others have shown no difference [40], similar to our results. In terms of aggressive behaviours, male pigs generally perform more aggression than females [41], and this could be particularly the case when males are not castrated.

5. Conclusions

This experiment demonstrated that it is possible to rear undocked pigs on fully slatted floors when a relatively large variety, quantity, and quality of slat-compatible enrichment is provided. Although mostly minor tail biting events occurred, the enrichment provided sustained pigs' interest and limited the escalation of tail biting towards an outbreak. This underlines the importance of environmental enrichment in keeping pigs sufficiently stimulated and in reducing harmful behaviours. Varying the types of enrichment provided over time did not significantly reduce tail damage. Similarly, no differences in damaging behaviours were observed, but pigs having the same enrichment interacted with the loose materials in the rack more. The study also showed that pigs have preferences for certain enrichment materials, with loose materials in a long rack being used the most, and therefore, it is important to consider enrichment characteristics, presentation, location, and maintenance when providing enrichment. The next step is to find a balance between the level of environmental enrichment that needs to be provided to reduce the risk of tail biting and the level that is economically feasible and practical to manage for farmers.

Author Contributions: Conceptualization, J.-Y.C., R.B.D.E., D.A.S., and K.O.D.; methodology, J.-Y.C., R.B.D.E., D.A.S., and K.O.D.; validation, R.B.D.E., D.A.S., and K.O.D.; formal analysis, J.-Y.C., K.O.D., and C.M.V.D.; investigation, C.M.V.D. and J.-Y.C.; writing—original draft preparation, C.M.V.D. and J.-Y.C.; writing—review and editing, R.B.D.E., D.A.S., and K.O.D.; visualization, J.-Y.C. and K.O.D.; supervision, R.B.D.E., D.A.S., and K.O.D.; project administration, R.B.D.E., D.A.S., and K.O.D.

Funding: This research received no external funding, but the authors in the study were supported by Teagasc and SRUC. Teagasc provided the funds for this open access publication. SRUC also received funding from the Rural and Environmental Science and Analytical Services Division of the Scottish Government.

Acknowledgments: We would like to thank the farm staff in Moorepark Pig Research Facility for their help in conducting the study and their care for the pigs, especially the farm manager Tomas Ryan's support in the purchase of most enrichment materials. We would also like to acknowledge the people and organisations that generously helped in providing free materials as enrichment items to facilitate this experiment, especially Michelle Liddane (Teagasc), John Finnan (Teagasc), and Milltech Digital Ltd.

Conflicts of Interest: The authors declare no conflict of interest.

Appendix A. Enrichment Scheduling

For the "SAME" pens, 1 out of 4 combinations was assigned at the start of weaner and finisher stage and stayed in the pen throughout the production stage. For the "SWITCH" pens, the combinations were switched every 2 weeks. Table A1 illustrates how each combination was equally represented in the SAME and SWITCH pens. The total week-in-use in all pens for each combination was also the same.

Table A1. The distribution of the enrichment combinations in each treatment group and pen.

Treatment/Pen	SAME				SWITCH			
Week	1	2	3	4	1	2	3	4
0	1	2	3	4	3	4	1	2
2	1	2	3	4	2	3	4	1
4	1	2	3	4	1	2	3	4
6	1	2	3	4	4	1	2	3
8 *	1	2	3	4	3	4	1	2

* Only in the finisher stage since the pigs spent 7 weeks in the weaner stage and 9 weeks in the finisher stage.

References

1. Sutherland, M.A.; Tucker, C.B. The long and short of it: A review of tail docking in farm animals. *Appl. Anim. Behav. Sci.* **2011**, *135*, 179–191. [CrossRef]
2. Harley, S.; Boyle, L.; O'Connell, N.; More, S.; Teixeira, D.; Hanlon, A. Docking the value of pigmeat? Prevalence and financial implications of welfare lesions in Irish slaughter pigs. *Anim. Welf.* **2014**, *23*, 275–285. [CrossRef]
3. Sonoda, L.T.; Fels, M.; Oczak, M.; Vranken, E.; Ismayilova, G.; Guarino, M.; Viazzi, S.; Bahr, C.; Berck, D. Tail Biting in pigs—Causes and management intervention strategies to reduce the behavioural disorder. A review. *Berl. Münch. Tierärztl. Wochenschr.* **2013**, *126*, 104–112. [PubMed]
4. Lahrmann, H.P.; Busch, M.E.; D'Eath, R.B.; Forkman, B.; Hansen, C.F. More tail lesions among undocked than tail docked pigs in a conventional herd. *Animal* **2017**, *11*, 1825–1831. [CrossRef] [PubMed]
5. European Union. *Council Directive 2008/120/EC of 18 December 2008 Laying Down Minimum Standards for the Protection of Pigs (Codified Version)*; European Union: Brussels, Belgium, 2009; p. L47/5-13.
6. Grümpel, A.; Krieter, J.; Veit, C.; Dippel, S. Factors influencing the risk for tail lesions in weaner pigs (Sus scrofa). *Livest. Sci.* **2018**, *216*, 219–226. [CrossRef]
7. Valros, A. Tail biting. In *Advances in Pig Welfare*; Spinka, M., Ed.; The Advances in Farm Animal Welfare series; Elsevier: Duxford, UK, 2018; pp. 137–166.
8. De Briyne, N.; Berg, C.; Blaha, T.; Palzer, A.; Temple, D. Phasing out pig tail docking in the EU—Present state, challenges and possibilities. *Porcine Health Manag.* **2018**, *4*, 27. [CrossRef] [PubMed]
9. Kallio, P.; Janczak, A.; Valros, A.; Edwards, S.; Heinonen, M. Case control study on environmental, nutritional and management-based risk factors for tail-biting in long-tailed pigs. *Anim. Welf.* **2018**, *27*, 21–34. [CrossRef]
10. Scollo, A.; Gottardo, F.; Contiero, B.; Edwards, S.A. A cross-sectional study for predicting tail biting risk in pig farms using classification and regression tree analysis. *Prev. Vet. Med.* **2017**, *146*, 114–120. [CrossRef]
11. D'Eath, R.B.; Niemi, J.K.; Vosough Ahmadi, B.; Rutherford, K.M.D.; Ison, S.H.; Turner, S.P.; Anker, H.T.; Jensen, T.; Busch, M.E.; Jensen, K.K.; et al. Why are most EU pigs tail docked? Economic and ethical analysis of four pig housing and management scenarios in the light of EU legislation and animal welfare outcomes. *Animal* **2015**, *10*, 687–699. [CrossRef]
12. Bulens, A.; Van Beirendonck, S.; Van Thielen, J.; Buys, N.; Driessen, B. Straw applications in growing pigs: Effects on behavior, straw use and growth. *Appl. Anim. Behav. Sci.* **2015**, *169*, 26–32. [CrossRef]
13. Chou, J.-Y.; D'Eath, R.B.; Sandercock, D.A.; Waran, N.; Haigh, A.; O'Driscoll, K. Use of different wood types as environmental enrichment to manage tail biting in docked pigs in a commercial fully-slatted system. *Livest. Sci.* **2018**, *213*, 19–27. [CrossRef]
14. Haigh, A.; Chou, J.-Y.; O'Driscoll, K. An investigation into the effectiveness of compressed straw blocks in reducing abnormal behaviour in growing pigs. *Animal* **2019**. Manuscript accepted for publication.

15. Holling, C.; grosse Beilage, E.; Vidondo, B.; Nathues, C. Provision of straw by a foraging tower –effect on tail biting in weaners and fattening pigs. *Porcine Health Manag.* **2017**, *3*. [CrossRef] [PubMed]
16. O'Driscoll, K.; McAuliffe, S.; Chou, J.-Y. Easyfix enrichment solutions for pig welfare. In Proceedings of the Banff Pork Seminar: Advances in Pork Production, Banff Pork Seminar, AB, Canada, 8–10 Jan 2019; Volume 30.
17. Veit, C.; Traulsen, I.; Hasler, M.; Tölle, K.H.; Burfeind, O.; grosse Beilage, E.; Krieter, J. Influence of raw material on the occurrence of tail-biting in undocked pigs. *Livest. Sci.* **2016**, *191*, 125–131. [CrossRef]
18. Zonderland, J.J.; Wolthuis-Fillerup, M.; Van Reenen, C.G.; Bracke, M.B.; Kemp, B.; Den Hartog, L.A.; Spoolder, H.A. Prevention and treatment of tail biting in weaned piglets. *Appl. Anim. Behav. Sci.* **2008**, *110*, 269–281. [CrossRef]
19. Van de Weerd, H.A.; Docking, C.M.; Day, J.E.L.; Avery, P.J.; Edwards, S.A. A systematic approach towards developing environmental enrichment for pigs. *Appl. Anim. Behav. Sci.* **2003**, *84*, 101–118. [CrossRef]
20. Van de Weerd, H.A.; Day, J.E.L. A review of environmental enrichment for pigs housed in intensive housing systems. *Appl. Anim. Behav. Sci.* **2009**, *116*, 1–20. [CrossRef]
21. Studnitz, M.; Jensen, M.B.; Pedersen, L.J. Why do pigs root and in what will they root? *Appl. Anim. Behav. Sci.* **2007**, *107*, 183–197. [CrossRef]
22. Trickett, S.L.; Guy, J.H.; Edwards, S.A. The role of novelty in environmental enrichment for the weaned pig. *Appl. Anim. Behav. Sci.* **2009**, *116*, 45–51. [CrossRef]
23. Chou, J.-Y.; D'Eath, R.B.; Sandercock, D.A.; O'Driscoll, K. Can increased dietary fibre level and a single enrichment device reduce the risk of tail biting in undocked pigs on fully slatted systems? **2019**. manuscript in preparation for submission to *Animal*.
24. Farewelldock Tail Scoring Protocol. Available online: http://farewelldock.eu (accessed on 26 January 2019).
25. DeBoer, S.; Garner, J.; McCain, R.; Lay, D., Jr.; Eicher, S.; Marchant-Forde, J. An initial investigation into the effects of isolation and enrichment on the welfare of laboratory pigs housed in the PigTurn® system, assessed using tear staining, behaviour, physiology and haematology. *Anim. Welf.* **2015**, *24*, 15–27. [CrossRef]
26. Harley, S.; More, S.J.; O'Connell, N.E.; Hanlon, A.; Teixeira, D.; Boyle, L. Evaluating the prevalence of tail biting and carcase condemnations in slaughter pigs in the Republic and Northern Ireland, and the potential of abattoir meat inspection as a welfare surveillance tool. *Vet. Rec.* **2012**, *171*, 621. [CrossRef] [PubMed]
27. Newberry, R.C.; Wood-Gush, D.G.M.; Hall, J.W. Playful behaviour of piglets. *Behav. Processes* **1988**, *17*, 205–216. [CrossRef]
28. Donaldson, T.M.; Newberry, R.C.; Špinka, M.; Cloutier, S. Effects of early play experience on play behaviour of piglets after weaning. *Appl. Anim. Behav. Sci.* **2002**, *79*, 221–231. [CrossRef]
29. Scott, K.; Taylor, L.; Gill, B.P.; Edwards, S.A. Influence of different types of environmental enrichment on the behaviour of finishing pigs in two different housing systems 2. Ratio of pigs to enrichment. *Appl. Anim. Behav. Sci.* **2007**, *105*, 51–58. [CrossRef]
30. Van de Weerd, H.A.; Docking, C.M.; Day, J.E.L.; Breuer, K.; Edwards, S.A. Effects of species-relevant environmental enrichment on the behaviour and productivity of finishing pigs. *Appl. Anim. Behav. Sci.* **2006**, *99*, 230–247. [CrossRef]
31. Scott, K.; Taylor, L.; Gill, B.P.; Edwards, S.A. Influence of different types of environmental enrichment on the behaviour of finishing pigs in two different housing systems 1. Hanging toy versus rootable substrate. *Appl. Anim. Behav. Sci.* **2006**, *99*, 222–229. [CrossRef]
32. Scott, K.; Taylor, L.; Gill, B.P.; Edwards, S.A. Influence of different types of environmental enrichment on the behaviour of finishing pigs in two different housing systems 3. Hanging toy versus rootable toy of the same material. *Appl. Anim. Behav. Sci.* **2009**, *116*, 186–190. [CrossRef]
33. Zwicker, B.; Gygax, L.; Wechsler, B.; Weber, R. Short- and long-term effects of eight enrichment materials on the behaviour of finishing pigs fed ad libitum or restrictively. *Appl. Anim. Behav. Sci.* **2013**, *144*, 31–38. [CrossRef]
34. Guy, J.H.; Meads, Z.A.; Shiel, R.S.; Edwards, S.A. The effect of combining different environmental enrichment materials on enrichment use by growing pigs. *Appl. Anim. Behav. Sci.* **2013**, *144*, 102–107. [CrossRef]
35. Telkänranta, H.; Bracke, M.B.M.; Valros, A. Fresh wood reduces tail and ear biting and increases exploratory behaviour in finishing pigs. *Appl. Anim. Behav. Sci.* **2014**, *161*, 51–59. [CrossRef]

36. Docking, C.M.; Van de Weerd, H.A.; Day, J.E.L.; Edwards, S.A. The influence of age on the use of potential enrichment objects and synchronisation of behaviour of pigs. *Appl. Anim. Behav. Sci.* **2008**, *110*, 244–257. [CrossRef]
37. Diana, A.; Manzanilla, E.G.; Calderón Díaz, J.A.; Leonard, F.C.; Boyle, L.A. Do weaner pigs need in-feed antibiotics to ensure good health and welfare? *PLoS ONE* **2017**, *12*, e0185622. [CrossRef] [PubMed]
38. Van Staaveren, N.; Calderón Díaz, J.A.; Garcia Manzanilla, E.; Hanlon, A.; Boyle, L.A. Prevalence of welfare outcomes in the weaner and finisher stages of the production cycle on 31 Irish pig farms. *Ir. Vet. J.* **2018**, *71*, 9. [CrossRef] [PubMed]
39. Schrøder-Petersen, D.; Simonsen, H. Tail Biting in Pigs. *Vet. J.* **2001**, *162*, 196–210. [CrossRef] [PubMed]
40. Taylor, N.R.; Main, D.C.J.; Mendl, M.; Edwards, S.A. Tail-biting: A new perspective. *Vet. J.* **2010**, *186*, 137–147. [CrossRef]
41. O'Driscoll, K.; O'Gorman, D.M.; Taylor, S.; Boyle, L.A. The influence of a magnesium-rich marine extract on behaviour, salivary cortisol levels and skin lesions in growing pigs. *Animal* **2012**, *7*, 1017–1027. [CrossRef]

© 2019 by the authors. Licensee MDPI, Basel, Switzerland. This article is an open access article distributed under the terms and conditions of the Creative Commons Attribution (CC BY) license (http://creativecommons.org/licenses/by/4.0/).

Article

The Use of Garlic Oil for Olfactory Enrichment Increases the Use of Ropes in Weaned Pigs

Nicola Blackie [1],* and Megan de Sousa [2]

[1] Pathobiology and Production Sciences, Animal Welfare Science and Ethics, Royal Veterinary College, Hawkshead Lane, Hatfield AL9 7TA, Hertfordshire, UK
[2] Centre for Equine and Animal Science, Writtle University College, Chelmsford CM1 3RR, Essex, UK; megandesousa@live.co.uk
* Correspondence: nblackie@rvc.ac.uk; Tel.: +44-(0)1707-666333

Received: 25 February 2019; Accepted: 3 April 2019; Published: 5 April 2019

Simple Summary: Pigs are highly intelligent and can be prone to tail biting behavior if their environment is lacking in complexity, which is a serious welfare concern. For disease control and hygiene, pigs are often kept in semi-barren environments to separate them from their faeces; this may include slatted floors. Slatted pens are also cheaper to maintain with straw being expensive in many countries. To make the environment less barren, pigs are required by law to have environmental enrichment or "toys". In this study, we designed a novel enrichment consisting of garlic-scented rope. We compared the pigs' current enrichment with either unscented or garlic-scented ropes for two weeks after weaning. We found that the pigs spent more time interacting with the garlic rope and that more pigs used it. Pigs also showed a preference for the rope with 84% of focal pigs choosing the garlic over the control rope. When the garlic ropes were re-sprayed on day 8, we saw an increase in the number of pigs using the garlic rope and the time spent interacting with it. This indicates that this caught their interest again as interactions had decreased over time. This might be useful to encourage pigs to use this enrichment.

Abstract: Pig producers are required to provide environmental enrichment to provide pigs the opportunity to perform investigative and manipulative behaviours (EU directive 2001/93/EC). Preventing enrichment from losing its novelty and decreasing the rate at which animals become habituated is important to maintain use of enrichment over time. A comparative study was formulated to identify whether weaner pigs housed in a semi-barren environment displayed a preference for olfactory enrichment compared to non-scented enrichment. Pigs ($n = 146$) were selected at 28 days old from two different batches ($n = 76$ and $n = 70$) and divided into pens. All pigs were given a control and a treatment (garlic scented) rope. Behavioural observations and rope interactions were assessed through direct observation. Throughout the entire study, the length of interaction with the garlic device was significantly higher ($p < 0.02$), indicating that there was a preference for olfactory enrichment compared to an odourless device. There was no significant occurrence of tail, ear, or flank biting in both batches. Weaner pigs showed a preference towards olfactory enrichment. Although habituation began to occur, this effect was mitigated by re-spraying the ropes, which resulted in increased interactions.

Keywords: Pig; enrichment; welfare; tail biting; post-weaning; garlic oil; olfactory

1. Introduction

Pigs within the EU must have access to environmental enrichment according to the EU directive 2001/93/EC. This enrichment must encourage pigs to undertake investigation and manipulation activities. The type of enrichment provided can influence how much the pigs use it. Those that stimulate pigs to forage and explore seem to keep pigs interested longer [1]. Pigs kept in an enriched environment post-weaning show more exploratory behaviours, less aggression, and appear to have a more positive affective state compared to those kept in barren environments [2,3]. Some enrichment types seem to be favoured in comparison to others, for example, if the enrichment has destructible, ingestible, or odorous properties, this may be more attractive to the pigs [1,4–6]. Enrichment devices that comprise of both olfactory and gustatory factors are highly important to pigs, due to their innate curiosity. Enrichment that supports rooting and foraging is also important as pigs kept in a semi-natural environment spend around half of their time foraging [7] and rooting has been shown to be a priority to pigs [6]. In a slatted environment, pigs may not have access to straw, which is noted to provide the best form of enrichment [8]. However, rope and other chewable substrates have been shown to be appropriate and that metal objects are not as suitable as an enrichment for pigs. Rope results in more interactions compared to chain, sawdust and shavings [9] and leads to more manipulation in finishing and early-weaned pigs compared to chain, pipe, and wood [10,11]. Flavoured ropes have been investigated to collect oral fluid from weaner and finishing pigs [12]; these ropes were soaked in sweet-citrus flavours (sucrose solution, apple juice, and pineapple juice). Sweet/citrus flavoured ropes did not increase oral fluid secretion and behaviour was not evaluated [12]. When plain (unflavoured) ropes were used to collect oral fluid, more that 80% of pigs chewed the ropes in the time that they were offered although this was influenced by housing, with straw based pigs taking longer to interact with the ropes and performing less chewing compared to slatted pigs [13]. This suggests that rope is a valuable enrichment item for pigs of all ages.

Pigs seem to be non-averse to garlic, with a study by Janz et al. [14] demonstrating that pigs offered garlic treated feed showed a preference through increased intake and liveweight gains. However, the levels of garlic inclusion might alter pigs' perception with some studies showing improvements in feed intake [14,15] with the addition of garlic and others showing no effects [16]. Garlic has strong olfactory properties and as such, the aim of this experiment was to investigate the impact of soaking rope in garlic oil as a novel enrichment device for weaned pigs. To date, there have been limited studies looking at olfactory enrichment in pigs.

2. Materials and Methods

The study was conducted at Sturgeons Farm, a commercial breeder-finisher unit part of Writtle University College, between December 2015 and January 2016. The Writtle University College Ethics Committee approved this research (98337238-AW1).

One hundred and fifty (Landrace X Large White) weaner pigs were recruited onto the study from two batches, the first containing 76 and the second 70. Four piglets had to be removed from the study due to ill health. Piglets were allocated to the pens at random with litters being mixed. All piglets were born in standard farrowing crates and management and nutrition of all sows and litters were equal; 3 rooms containing 4 crates were used for each batch. Piglets had access to a heated creep area and were offered creep feed prior to weaning (Ultima 1 Starter Feed). At birth, piglets were cross-fostered as required and received 1 mL of iron within the first 24 h. Tail and teeth reduction were both performed on all piglets as approved by the veterinary surgeon at the quarterly visit. The pigs were 4 weeks old at weaning and were housed in pens (25/pen, 6 pens over 2 replicates) according to weight. All pens measured 250 cm by 350 cm and had slatted plastic flooring. Each pen consisted of a single trough where feeding occurred at around 8:30 am and 5:00 pm to allow ad lib feeding, this coincided with the turning on and off of lights. Each trough allowed roughly four weaners to feed at one single time although this did not appear to cause aggression. The entire barn comprised of a fan assisted ventilation system, with a solitary fan in each pen leading to the exterior. Temperature in the barn rarely fluctuated; newly weaned pigs received a direct heat source to compensate for the lack of bedding.

Pens had two close-knit chains of a 6 mm diameter and 1.8 m in length attached to a metal bracket with an equal distance of 45 cm apart. Attached to these chains were either a control (unscented) rope or treatment (garlic) rope, which were soaked in a concentrated garlic oil extract (Pegasus Health) mixed with purified water. This solution was used in a ratio of 30 mL (oil): 1 L (water) ropes were soaked individually for a 4-h period and dried overnight. All ropes were then attached 2 h before the pigs initially entered the pen. In the first batch of each pen, the control rope was attached to the right chain and the garlic flavoured to the left, then reversed in the second batch. The rope used in this study was natural cotton rope, measuring 50 cm in length with a diameter of 30 mm. Ropes were tied at the top, with the remainder of the fibres split into 3 individual ropes, thus allowing fraying at the base. The pigs had a standard piece of enrichment in addition, which was a porcichew attached to a chain. Garlic ropes were re-sprayed with a concentration of 10 mL:500 mL on day 8 of the trial and allowed to dry for 30 min before being returned to the pen.

Behavioural observations were undertaken and consisted of focal observations. Ten focal pigs per batch were selected at random and marked with a stock marker to give individual identification; equal numbers of males and females were selected. Instantaneous focal sampling of behaviour was carried out at 2 min intervals for 1 h (n = 30 observations per hour observation period). The recorded behaviours were mutually exclusive and are detailed in the ethogram (Table 1) and all pigs were observed. Focal pigs were used to describe preferences for rope.

Table 1. Ethogram of behaviours recorded in weaned pigs.

Behaviour	Description
Standing	Pig stood on all four limbs stationary
Feeding/drinking	Pig consuming supplied feed from trough/Pig latched on drinker and swallowing water
Interacting with others	Pigs actively seeking other pen-mate and chasing, sniffing, belly nosing and other general interactions
Enrichment use	Interaction the ropes including sniffing, chewing, thrashing with rope in mouth or walking with rope
Exploring environment	Rooting or licking action around parameter of environment
Locomotion	Pig walking around the pen with no clear purpose other than to reach a different area/device
Aggression (fighting)	Individuals nosing, biting resulting in both parties rearing onto hind limbs and pushing each other until an individual has fallen
Tail/ear/flank, biting	Individual seeking out aforementioned appendage and biting
Resting/sleeping	Animal lying down with eyes open or closed
O.F.S. (out of sight)	Selected individual is not visible

Measurements of pigs interacting with ropes were collated by recording the length of interaction (start and stop time of interaction), number of individuals interacting, and the frequency of the interaction with each rope. One observer (MdS) spent an hour a day observing enrichment use, equating to 14 h of observations per pen of pigs over 2 weeks. This observation time varied to cover as much of the day as possible. To avoid an effect of the observer on pig behaviour, a habituation period of 15 to 20 min was allowed before each observation period. Rope interactions were collated from all pigs within each pen, however, animal preference was documented from the focal pigs in each pen.

All data were tested for normality using the Pearson's skewness test prior to analysis. From the scan sampling, the number of pigs interacting with the ropes within the hour of observation were added up for each day. These data were then analysed using a repeated measure analysis of variance with the repeated measure being the 14 days. For this test, the factors assessed were the effects of treatment, time, and the interaction between treatment and time. For the time spent interacting with the ropes, this was calculated from the focal pigs and this was also assessed using a repeated measures analysis of variance. The 30 observations per day were summed and these data were used.

Time budgets and behaviours were statistically analysed using Kruskal-Wallis to identify the most performed behaviour alongside a Mann-Whitney U test as a post hoc test. All statistical analyses were completed through GenStat version 17 (VSN International Ltd., Hemel Hempstead, UK).

3. Results

Differences were seen in the number and length of rope interactions when comparing the treatment (garlic treated rope) and control (untreated rope).

3.1. Activity Budget of Piglets

The most performed behaviour across both batches and all pens equalled 'resting/sleeping' ($p < 0.001$), indicating that weaner pigs spend the majority of their day asleep or resting (23.7% of observations). The next most prevalent behaviour of the focal pigs was interaction with the garlic rope, which made up 17.2% of their time ($p < 0.002$). Pigs also spent 14% of their time interacting with others and 4.3% of their time interacting with the control rope. Exploring the environment was undertaken in 12.1% of observations ($p < 0.005$). Upon re-spraying the garlic extract, an increase of the time spent interacting with the garlic rope was noted, with a 17.5% increase of time spent interacting was observed. There were no significant incidences of injurious behaviours performed throughout the study.

3.2. Rope Interactions

Soaking rope in garlic oil had an effect on the length of time individuals spent interacting with the rope compared to the control (unscented) enrichment device ($p < 0.02$, Figure 1). There was a rise in interaction with the garlic rope at the point of re-spray as seen in day 8 in Figure 1. At the point of re-spraying the garlic extract, the interaction levels increased compared to the day before re-spraying, a 73% increase in activity was recorded.

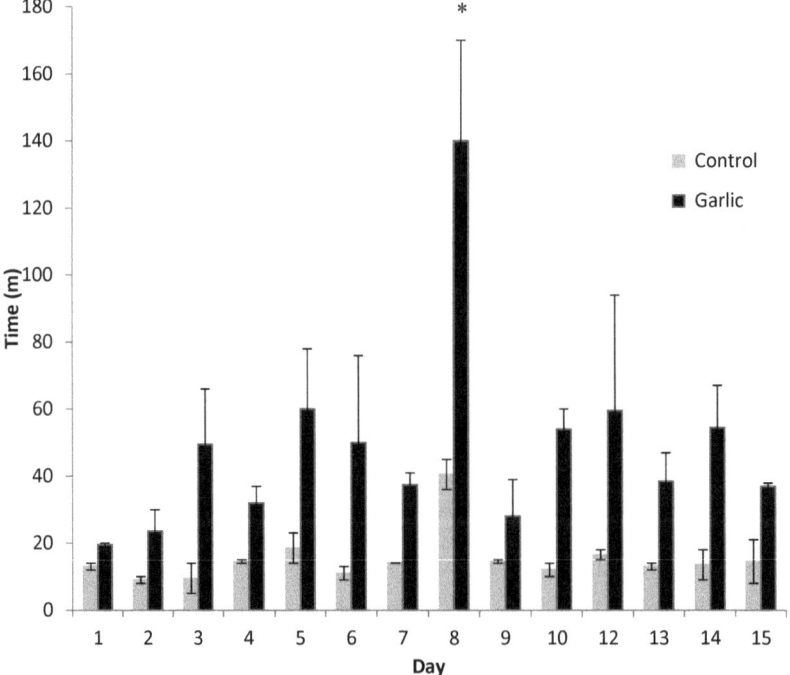

Figure 1. Length of time spent interacting with the control (untreated) rope compared with the garlic treated rope for each day over the study period. There was a significant effect of treatment ($p = 0.017$), but no effect of time ($p = 0.159$) or a treatment x time interaction ($p = 0.307$). The asterisk represents when the treatment rope was re-sprayed with garlic oil (day 8).

With regards to the number of individual pigs interacting with the ropes, there were significantly more pigs interacting with the garlic treated rope ($p = 0.004$) within a one hour period of observation (Figure 2). The number of pigs interacting with the ropes showed a similar pattern to the amount of time spent interacting with the ropes. Again, an increase in the number of interactions was observed on the day on which the rope was re-sprayed (day 8). The number of pigs interacting increased by 57% in comparison to the previous day. Figure 3 shows the pigs using the rope enrichment; the garlic soaked rope is on the left and the control rope is on the right of the image.

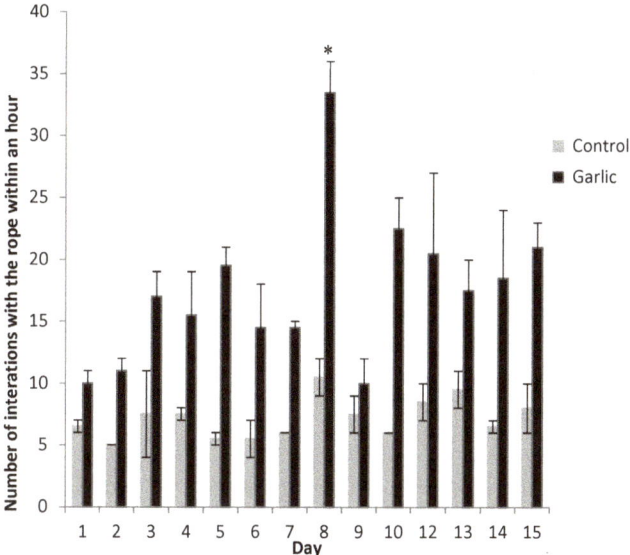

Figure 2. Number of interactions observed within an hour for the control (untreated) rope compared with the garlic treated rope each day over the study period. There was a significant effect of treatment ($p = 0.004$), but no effect of time ($p = 0.143$) or a treatment x time interaction ($p = 0.242$). The asterisk represents when the treatment rope was re-sprayed with garlic oil (day 8).

Figure 3. An example of pigs interacting with the garlic rope (left) or control rope (right). Focal pigs were marked with a spray marker to give individual identification.

Throughout the entire study, focal animals preferred the garlic rope by using the ropes for a greater amount of time. Individuals interacted with the garlic rope for an increased amount of time

(>70% of individuals showed a garlic rope preference) compared to the control treatment. When focal pigs were seen interacting with the enrichment, 84% used the garlic rope and 16% used the control.

4. Discussion

Given the popularity of the garlic scented rope expressed in both the time interacting with it and the number of pigs interacting with it, it would appear that it is an effective olfactory enrichment for pigs. Olfactory ques are used by pigs to recognize familiar conspecifics [17], locate the udder [18], and play an important part in feed palatability [19]. Piglets showed a preference towards olfactory enrichment and spent a lot of time of their day interacting with the device, distracting them from performing any injurious behaviours, which were not observed. Pigs in the present study were kept in a barren environment and spent proportionally more time interacting with the garlic rope and the environment than interacting with each other. Our study did not capture oral fluid from ropes like other studies [12,13], however, the use of garlic oil could be investigated to see if it improves fluid recovery.

Habituation is described as the process of an object losing novelty and an animal no longer responding to an object's presence [5]. Once an animal becomes habituated to a stimulus, the motivation to interact with the stimulus lessens. If the object is not initially attractive to the animal, interest will decrease quickly [5,20], thus leading to a potentially enriched environment becoming barren. Habituation occurs even when enrichment is rotated or changed [5,9], although the number of pigs interacting with each device began to decrease over time until the re-application of garlic oil; the interaction with garlic was always significantly higher than that of the control and other enrichment devices. Habituation may have taken place as is often seen with enrichment devices, however, once garlic ropes were re-sprayed, use of the devices increased by 35% in total. Interaction activity remained at an increased level post spraying for three days across all results at a level of 46%, supporting the theory of Van de Weerd et al. [1] of applying an odour to an enrichment device to reduce habituation. Length of interaction with the garlic rope increased following respraying, i.e., increased from 40 min/day to 140 min/day as well as the number of pigs using it. Interestingly, the control ropes were also manipulated more after the garlic ropes were refreshed. This may be facilitation behaviour or simply that not all the pigs could interact with the garlic rope at one time. Pigs who may want to interact with the garlic rope, but could not due to lack of space may have chosen to manipulate the control rope instead.

This re-interest in the enrichment may be of particular importance in the event of a tail biting outbreak and it has been shown that extra enrichment can act as an intervention against tail biting [21]. The pigs in this study were docked to avoid outbreaks of tail biting and this is recognised as being in the pigs best interest [22]. We would like to repeat the study on undocked pigs to investigate the impact of our enrichment strategy on tail biting.

The method of preference testing in this study was a simple choice test, rather than testing based on motivation and drive. This is mainly due to the nature of the intensive pig production system and the restrictions of altering the running of the unit. As motivation tests often rely on the animal experiencing some sort of cost to access the device [23], this is not ideal in a farming environment. Choice method testing proved successful in this study, as the aim was to identify whether pigs have a preference for olfactory enrichment over unscented rather than assessing how strong the motivation is to access the different enrichments. Although some of these methods do overlap, it is generally acknowledged that if an animal spends a greater quantity of time interacting with a certain device instead of another, then the animal prefers the primary stimulus [24]. It is argued that the resource the animal interacts with more is not necessarily preferred; however, this is unlikely in the present study due to a lack of fluctuations in the environment, social grouping, and fixtures and fittings [25]. An issue of laterality could possibly be raised in this study, as ropes scented or unscented were placed on the right or left. Farmer and Christison [26] stated that piglets and weaners did not show any laterality towards types of flooring as behaviours were equally performed regardless of whether specific flooring was located at the left or the right of the animal. While laterality research in pigs is limited, it is unlikely that laterality was present in this study and in fact, a preference for the garlic device was shown regardless of which side the treatments were offered.

5. Conclusions

The results from this study suggest that weaner pigs prefer olfactory enrichment to non-olfactory enrichment. From these results, it can be concluded that garlic is an attractive odour and flavour to juvenile pigs. When the ropes were re-sprayed, the novelty increased, suggesting the addition of scent reduces habituation to an enrichment device. Garlic scented ropes may be useful for distracting pigs when a tail-biting outbreak occurs.

Author Contributions: Conceptualization, N.B. and M.d.S.; Data curation, M.d.S.; Formal analysis, M.d.S.; Investigation, M.d.S.; Methodology, N.B. and M.d.S.; Supervision, N.B.; Writing—original draft, N.B.; Writing—review & editing, M.d.S.

Funding: This research received no external funding.

Acknowledgments: The authors gratefully acknowledge help and support from Sturgeons farm and technical staff at Writtle University College.

Conflicts of Interest: The authors declare no conflict of interest.

References

1. Van de Weerd, H.A.; Docking, C.M.; Day, J.E.L.; Avery, P.J.; Edwards, S.A. A systematic approach towards developing environmental enrichment for pigs. *Appl. Anim. Behav. Sci.* **2003**, *84*, 101–118. [CrossRef]
2. Douglas, C.; Bateson, M.; Walsh, C.; Bédué, A.; Edwards, S.A. Environmental enrichment induces optimistic cognitive biases in pigs. *Appl. Anim. Behav. Sci.* **2012**, *139*, 65–73. [CrossRef]
3. Melotti, L.; Oostindjer, M.; Bolhuis, J.E.; Held, S.; Mendl, M. Coping personality type and environmental enrichment affect aggression at weaning in pigs. *Appl. Anim. Behav. Sci.* **2011**, *133*, 144–153. [CrossRef]
4. Feddes, J.J.R.; Fraser, D. Non-nutritive Chewing by Pigs: Implications for Tail-biting and Behavioral Enrichment. *Trans. ASAE* **1994**, *37*, 947–950. [CrossRef]
5. Trickett, S.L.; Guy, J.H.; Edwards, S.A. The role of novelty in environmental enrichment for the weaned pig. *Appl. Anim. Behav. Sci.* **2009**, *116*, 45–51. [CrossRef]
6. Studnitz, M.; Jensen, M.B.; Pedersen, L.J. Why do pigs root and in what will they root?: A review on the exploratory behaviour of pigs in relation to environmental enrichment. *Appl. Anim. Behav. Sci.* **2007**, *107*, 183–197. [CrossRef]
7. Stolba, A.; Wood-Gush, D.G.M. The behaviour of pigs in a semi-natural environment. *Anim. Sci.* **1989**, *48*, 419–425. [CrossRef]
8. Bracke, M.B.M.; Zonderland, J.J.; Lenskens, P.; Schouten, W.G.P.; Vermeer, H.; Spoolder, H.A.M.; Hendriks, H.J.M.; Hopster, H. Formalised review of environmental enrichment for pigs in relation to political decision making. *Appl. Anim. Behav. Sci.* **2006**, *98*, 165–182. [CrossRef]
9. Guy, J.H.; Meads, Z.A.; Shiel, R.S.; Edwards, S.A. The effect of combining different environmental enrichment materials on enrichment use by growing pigs. *Appl. Anim. Behav. Sci.* **2013**, *144*, 102–107. [CrossRef]
10. Zonderland, J.J.; Vermeer, H.M.; Vereijken, P.F.G.; Spoolder, H.A.M. Measuring a Pig's Preference for Suspended Toys by Using an Automated Recording Technique. *Agric. Eng. Int.* **2003**, *5*, 1–11.
11. Horrell, I.; Ness, P.A. Enrichment satisfying specific behavioural needs in early-weaned pigs. *Appl. Anim. Behav. Sci.* **1995**, *44*, 264. [CrossRef]
12. Dawson, L.L.; Edwards, S.A. The effects of flavored rope additives on commercial pen-based oral fluid yield in pigs. *J. Vet. Behav.* **2015**, *10*, 267–271. [CrossRef]
13. Seddon, Y.M.; Guy, J.H.; Edwards, S.A. Optimising oral fluid collection from groups of pigs: Effect of housing system and provision of ropes. *Vet. J.* **2012**, *193*, 180–184. [CrossRef] [PubMed]
14. Janz, J.A.M.; Morel, P.C.H.; Wilkinson, B.H.P.; Purchas, R.W. Preliminary investigation of the effects of low-level dietary inclusion of fragrant essential oils and oleoresins on pig performance and pork quality. *Meat Sci.* **2007**, *75*, 350–355. [CrossRef] [PubMed]

15. Langendijk, P.; Bolhuis, J.E.; Laurenssen, B.F.A. Effects of pre- and postnatal exposure to garlic and aniseed flavour on pre- and postweaning feed intake in pigs. *Livest. Sci.* **2007**, *108*, 284–287. [CrossRef]
16. Horton, G.M.J.; Blethen, D.B.; Prasad, B.M. The effect of garlic (*Allium sativum*) on feed palatability of horses and feed consumption, selected performance and blood parameters in sheep and swine. *Can. J. Anim. Sci.* **1991**, *71*, 607–610. [CrossRef]
17. Kristensen, H.H.; Jones, R.B.; Schofield, C.P.; White, R.P.; Wathes, C.M. The use of olfactory and other cues for social recognition by juvenile pigs. *Appl. Anim. Behav. Sci.* **2001**, *72*, 321–333. [CrossRef]
18. Morrow-Tesch, J.; McGlone, J.J. Sensory systems and nipple attachment behavior in neonatal pigs. *Physiol. Behav.* **1990**, *47*, 1–4. [CrossRef]
19. Roura, E.; Fu, M. Taste, nutrient sensing and feed intake in pigs (130 years of research: then, now and future). *Anim. Feed Sci. Technol.* **2017**, *233*, 3–12. [CrossRef]
20. Blackshaw, J.K.; Thomas, F.J.; Lee, J.-A. The effect of a fixed or free toy on the growth rate and aggressive behaviour of weaned pigs and the influence of hierarchy on initial investigation of the toys. *Appl. Anim. Behav. Sci.* **1997**, *53*, 203–212. [CrossRef]
21. Lahrmann, H.P.; Hansen, C.F.; D'Eath, R.B.; Busch, M.E.; Nielsen, J.P.; Forkman, B. Early intervention with enrichment can prevent tail biting outbreaks in weaner pigs. *Livest. Sci.* **2018**, *214*, 272–277. [CrossRef]
22. D'Eath, R.B.; Niemi, J.K.; Vosough Ahmadi, B.; Rutherford, K.M.D.; Ison, S.H.; Turner, S.P.; Anker, H.T.; Jensen, T.; Busch, M.E.; Jensen, K.K.; et al. Why are most EU pigs tail docked? Economic and ethical analysis of four pig housing and management scenarios in the light of EU legislation and animal welfare outcomes. *Animal* **2015**, *10*, 687–699. [CrossRef]
23. Fraser, D.; Matthews, L.R. Preference and Motivation Testing. In *Animal Welfare*; Appleby, M.C., Hughes, B.O., Eds.; CAB International: New York, NY, USA, 1997; pp. 159–173.
24. Rushen, J. Using aversion learning techniques to assess the mental state, suffering, and welfare of farm animals. *J. Anim. Sci.* **1996**, *74*, 1990–1995. [CrossRef] [PubMed]
25. Kirkden, R.D.; Pajor, E.A. Using preference, motivation and aversion tests to ask scientific questions about animals' feelings. *Appl. Anim. Behav. Sci.* **2006**, *100*, 29–47. [CrossRef]
26. Farmer, C.; Christison, G.I. Selection of Perforated Floors by Newborn and Weanling Pigs. *Can. J. Anim. Sci.* **1982**, *62*, 1229–1236. [CrossRef]

© 2019 by the authors. Licensee MDPI, Basel, Switzerland. This article is an open access article distributed under the terms and conditions of the Creative Commons Attribution (CC BY) license (http://creativecommons.org/licenses/by/4.0/).

Article

Effects of Enrichment Type, Presentation and Social Status on Enrichment Use and Behaviour of Sows with Electronic Sow Feeding

Cyril Roy [1], Lindsey Lippens [2], Victoria Kyeiwaa [1], Yolande M. Seddon [3], Laurie M. Connor [2] and Jennifer A. Brown [1,*]

1. Prairie Swine Centre, Box 21057, 2105-8th Street East, Saskatoon, SK S7H 5N9, Canada; Cyril.Roy@usask.ca (C.R.); Victoria.Kyeiwaa@usask.ca (V.K.)
2. Department of Animal Science, Faculty of Agricultural and Food Sciences, 201-12 Dafoe Road, University of Manitoba, Winnipeg, MB R3T 2N2, Canada; Lindsey_2310@hotmail.com (L.L.); Laurie.Connor@umanitoba.ca (L.M.C.)
3. Western College of Veterinary Medicine, University of Saskatchewan, 52 Campus Drive, Saskatoon, SK S7N 5B4, Canada; yolande.seddon@usask.ca
* Correspondence: jennifer.brown@usask.ca; Tel.: +1-306-667-7442

Received: 9 March 2019; Accepted: 7 June 2019; Published: 18 June 2019

Simple Summary: Many North American pork producers are transitioning to group housing systems for gestating sows and are looking to provide animals with environmental enrichment. Because straw poses biosecurity and manure management concerns it is important to identify alternative enrichments which can benefit sow welfare. The effects of four enrichment treatments were studied: (1) Constant: constant provision of wood on chain; (2) Rotate: rotation of three enrichments (rope, straw and wood on chain); (3) Stimulus: rotation of three enrichments with an associative stimulus (bell or whistle); and (4) Control: no enrichment. Contacts with enrichment and time spent in different postures were measured using scan sampling for all sows. Skin lesions were scored and cortisol was measured in saliva samples in a subset of dominant and subordinate sows. Sows spent more time contacting enrichment in Rotate and Stimulus treatments than Constant, particularly when straw was provided. Subordinate sows spent more time near enrichments, and more time standing than dominant sows. Subordinate sows also received more skin lesions and had higher salivary cortisol concentrations than dominants. We conclude that enrichments are valued by sows, and that the right amount is needed to minimize competition over access. Additional work is needed on rotational schedules and determining appropriate levels of enrichment for sows.

Abstract: The goal of this study was to identify practical enrichments for sows in partially or fully slatted pen systems. Four treatments were applied: (1) Constant: constant provision of wood on chain; (2) Rotate: rotation of rope, straw and wood enrichments; (3) Stimulus: rotation of enrichments (as in Rotate) with an associative stimulus (bell or whistle); and (4) Control: no enrichment, with each treatment lasting 12 days. Six groups of 20 ± 2 sows were studied from weeks 6 to 14 of gestation in pens with one electronic sow feeder. Each group received all treatments in random order. Six focal animals (3 dominant and 3 subordinate) were selected per pen using a feed competition test. Digital photos were collected at 10 min intervals for 8 h (between 8 a.m. and 4 p.m.) on 4 days/treatment (d 1, 8, 10 and 12) to record interactions with enrichment. Skin lesions were assessed on days 1 and 12, and saliva cortisol samples collected in weeks 6, 10 and 14 of gestation on focal pigs. Sows spent more time in contact with enrichments in Rotate and Stimulus treatments than Constant. Enrichment treatments did not influence lesion scores. Subordinate sows spent more time standing and near enrichments than dominants. Subordinate sows also received more skin lesions and had higher salivary cortisol concentrations than dominants. These results indicate that access to enrichment is valued by sows but can result in greater aggression directed towards subordinates.

Keywords: environmental enrichment; social status; sows; aggression; habituation

---

1. Introduction

As gestation stalls are replaced with group housing in North America, there are increasing opportunities to provide enrichment to sows and potential to improve the health and welfare of pigs [1]. Sows managed in group housing may benefit from social enrichment (interaction with other sows), if appropriate space allowance and feeding management are provided. However, particularly for subordinate group members, there are risks associated with group housing such as increased aggression [2,3] and potential health risks such as lameness, injury and abortion [4].

Scientific evidence on environmental enrichment for pigs indicates that it has the potential to increase social interactions, reduce stereotypies and increase general activity [5,6]. While research on enrichment use in sows is limited, some positive effects associated with enrichment in growing pigs include increased ability to adapt to novel situations, reduced incidence of negative behaviours such as belly nosing, tail and ear biting, reduced fearfulness and improved learning capabilities [1,7–12]. Extrapolating from these studies, it is hypothesised that some of the negative aspects of housing sows in groups, such as aggression and oral stereotypies can be mitigated by providing enrichment.

In North America, partially slatted or fully slatted floors with liquid manure systems are common and limit the use of substrate enrichments due to the potential for blocking the slurry system [10]. Other concerns related to substrate provision include cost and biosecurity risk. Point source object enrichment (enrichment provided at a fixed position in the pen) may be more suitable for slatted or partially slated flooring systems. Some point source enrichments that have been previously studied include a rubber bar, rubber ball, ribbon, rope, garden hose, wood, chains [8,13,14] straw or hay [1,7] and sexual pheromones [15].

Many factors are known to affect enrichment use by pigs. For enrichment to be effective, it should be kept clean and not soiled with feces, easily accessible, attractive to the animals, deformable, destructible, and immobile so that the animals are able to hold and manipulate it [13]. Research has shown that presenting enrichment materials in different ways also helps to maintain their attractiveness [16]. In piglets, announcing the provision of enrichment (by using a sound stimulus) resulted in a significant increase in play behaviour and reductions in aggression and lesion scores following weaning [9]. Social status within the group can also affect activity and the behaviour of sows. If enrichment is valued and access is limited then dominant sows may displace subordinates and limit access to enrichment [17]. Additional research is needed to understand how these factors affect enrichment usage in sows.

Habituation can also reduce the effectiveness of enrichments, particularly when point source physical enrichments are provided [8,18]. When the same enrichment material is used over time, animals lose interest due to habituation [19]. The initial response and interaction with point source objects is generally high (within the first 24 h), and then declines over subsequent days or weeks [15]. However, habituation to a point source object is also dependent on its properties and the method of presentation [15]. For example, Van de Weerd et al. [18] observed that hanging sisal rope had a higher initial response and interaction frequency on day 1 when compared with loose concrete blocks or a rubber boot provided on the same day. This highlights the importance of understanding how different presentation methods affect enrichment use. Previous studies also indicate that pigs are attracted to enrichments that are destructible, deformable and chewable [8,13,18] indicating that the properties of the enrichment can also impact their use. Therefore, there is a need to understand how different types of enrichment and presentation methods influence habituation and enrichment usage by sows in group housing.

In this study, three enrichment treatments were provided to group housed sows and were compared against a control treatment involving no enrichment. Three enrichment materials, including cotton rope, wood and straw were selected for the trial. Our objective was to examine how enrichment type, method of presentation and the social rank of sows affected enrichment use and sow behaviour. We hypothesized that frequent rotation of enrichments would increase enrichment use and reduce habituation, and that the use of an associative stimulus would increase the initial response to enrichment provision. Furthermore, we hypothesized that enrichment provision would reduce social stress and aggression among sows.

2. Materials and Methods

The research was conducted at the research barn associated with Glen Lea Research Centre (University of Manitoba, Winnipeg, MB, Canada). The experiment was approved by University of Manitoba Committee on Animal Care (AUP# F2014-031/1/2) and adhered to the Canadian Council on Animal Care guidelines for humane animal use.

2.1. Animals and Housing

A total of 120 sows (20 ± 2 sows in 6 replicates, Genesus genetics, Oakville, Manitoba, Canada) were studied. Sows were housed in groups on partially slatted concrete floors. Trials were conducted in four gestation pens of two designs, with an average pen area of 61.2 m^2 (3 m^2/sow). Two pens had 25% solid floor area and two pens had 43% solid floor area, with the remaining portion being slatted (Figure 1). Feed was supplied using Electronic Sow Feeding (ESF; Nedap Velos, Nedap Livestock Management, Groenlo, The Netherlands). Sows were artificially inseminated with pooled semen, and remained in groups of 3–5 sows with feeding stalls for four to five weeks before being mixed into gestation pens after confirmation of pregnancy. Sows entered the study following mixing into gestation pens in week 4–5 of gestation.

A sub-sample of 3 dominant and 3 subordinate sows per group was identified and marked in week 1, based on feed competition trials, following procedures adapted from Anderson et al. [19]. Feed competition testing was performed in the pen on days 3–5 after mixing, after the initial aggression associated with mixing had resolved. On the afternoon of the third day after mixing, the solid floor area was scraped and sows were given 4 kg of feed poured onto the floor in two lines (2 kg per line, approx. 1 m long in 2 lines, spaced at least 2 m apart). Once the feed was placed, sows were allowed to compete for floor feed. This procedure was repeated on days 4 and 5 after mixing. On the third day of the feed competition trial, sows were observed and three sows which gained first access to feed were identified as 'dominant' (Dom), and three sows that refrained from feed competition and/or were driven away were identified as 'subordinate' (Sub).

2.2. Treatments

Four enrichment treatments were provided to each group of sows (pen). Each treatment was provided for 12 days, with treatment order randomized and a two-day interval between treatments. Treatments consisted of:

1. Constant provision of one type of enrichment—wood on chains, 3 per pen (Constant),
2. Rotation of three enrichments—rope, straw, wood on chain (Figure 2), 3 per pen (Rotate),
3. Rotation of three enrichments (as described for Rotate) with an associative stimulus used to signal the arrival of enrichment (Stimulus). The associative stimulus used was a bell or whistle (duration: 2 s), and was switched half-way through the study so that any sows that returned to the study (in their next gestation) would not be familiar with the stimulus, and
4. No enrichment (Control).

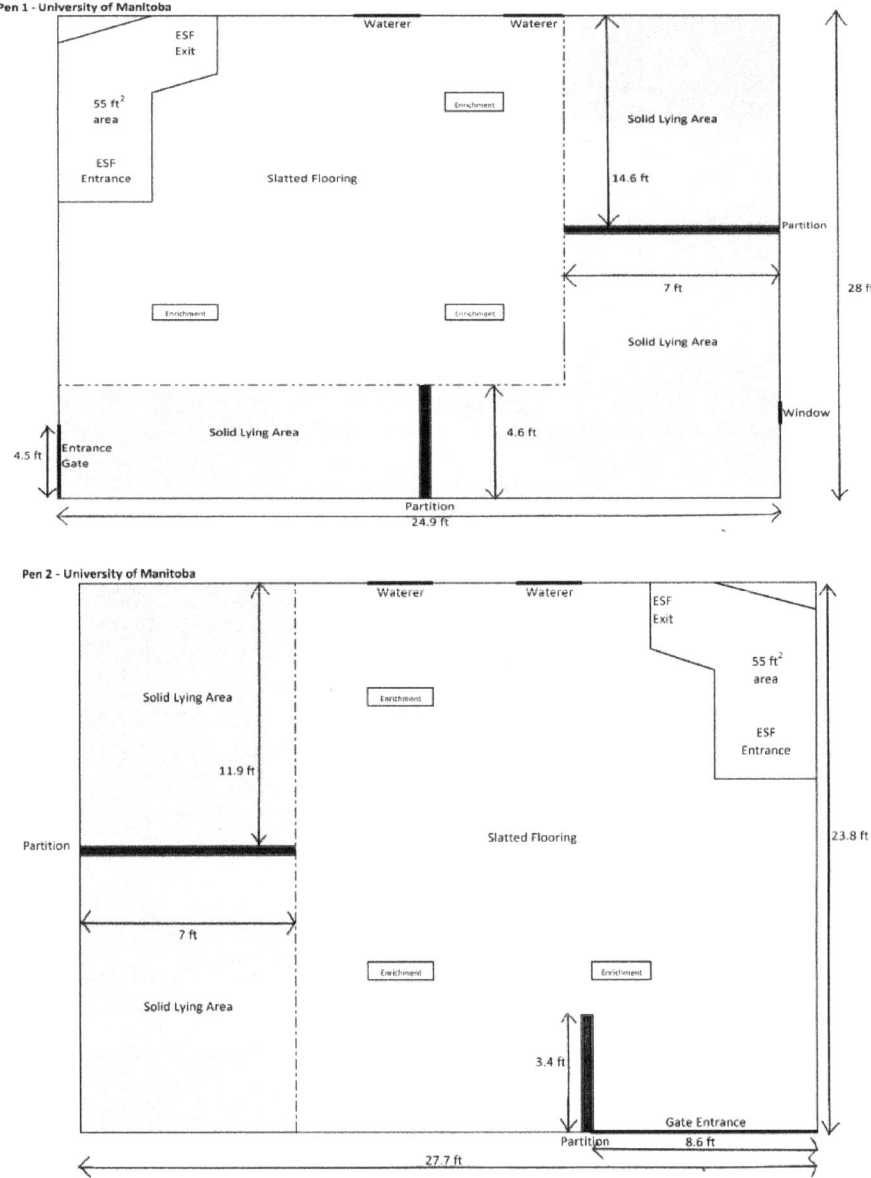

Figure 1. Pen designs used in the trial. Two of the pens had 43% solid floor area (**Pen 1**) and two pens had 25% solid floor area (**Pen 2**).

In Rotate and Stimulus treatments, the enrichments were rotated three times per week, with rope on Mondays, straw on Wednesdays and wood on Fridays (see timeline, Figure 3). For the Stimulus treatment, a bell or whistle was used immediately before providing enrichment as an associative stimulus. Wood on chain and rope enrichments were hung over the slatted area of each pen using carabiner clips to attach the enrichment to chains attached to the ceiling using an eye bolt. The rope length was 1.2 m (including a 15 cm tassel at the end), using three-stranded cotton rope 19 mm in diameter and suspended 20 cm above floor level. Straw was provided on the solid floor area such that there was 300 g per sow (8.4 kg in total). Any remaining straw was removed before giving the next enrichment, but most of the material was consumed by sows. The wood enrichment was made of softwood, 5 × 10 cm and 1.2 m in length and hung from the chain in such a way that it rested on the floor at 45° angle.

Figure 2. Photos illustrating presentation of straw (**A**), wood (**B**) and rope (**C**) enrichments to sows.

Day:	1	2	3	4	5	6	7	8	9	10	11	12
Enrichment:	Rope		Straw		Wood			Rope		Straw		Wood
Data collection:	Photos Lesions Weigh							Photos		Photos		Photos Lesions Weigh
			ST							ST		

Figure 3. Timeline illustrating data collection and the rotation schedule for enrichments in Rotation and Stimulus treatments. Each treatment lasted 12 days, followed by two days off. Four treatments were provided consecutively to each pen group ($n = 6$) in random order over a period of eight weeks (beginning at 5–6 weeks and ending at 13–14 weeks of gestation).

2.3. Data Collection

The study began the week that sows were moved to gestation pens (4–5 weeks of gestation). The body weight and parity of individual sows was recorded as sows entered group gestation pens.

Enrichment use was studied by mounting one camera over each pen and recording sow activity around the enrichments. Digital photos were taken at 10 min intervals over 8 h per day (8 a.m.–4 p.m.) on days 1, 8, 10 and 12 of each treatment (approximately 48 photos per day and 192 over the four days of observation). Photos were transcribed by one trained observer, who determined the location and posture of all sows visible in the enrichment area. All sows that were clearly visible were observed and the number of sows standing, lying or sitting at each time point was recorded. Table 1 shows an ethogram of behaviour categories recorded. For Rotation and Stimulus treatments, the recordings coincided with the first 8 h following provision of new enrichments. In the Stimulus treatment the "stimulus" (2 s, bell or whistle) was given as soon as cameras were started, and was followed by enrichment provision. Enrichment use was studied by transcribing the photos in two datasheets: First as a whole group and second recording only the 6 selected Dom and Sub sows. Dom and Sub sows were identified by blue (Dom) or red (Sub) spray markings.

Table 1. Ethogram used to identify sow posture and location relative to enrichment.

Location/Posture	Definition of Location or Posture
In contact	Sow's snout is in contact with enrichment. For straw enrichment, the sow should be standing on or facing the straw, and appears to be contacting the material with her snout.
Less than 1 m	Sow is not in contact with enrichment, but head is within approximately 1 m of enrichment. For straw enrichment, the sow is lying or standing in proximity to the straw (head is <1 m from straw)
Greater than 1 m	Sow's head is greater than 1 m from enrichment. Includes all visible sows in the photo that are not in contact or <1 m.
Standing	Sow is upright on four legs (not sitting or lying). If the sow is in the process of lying down, however still upright and rear end is supported on hind legs, she is considered standing.
Sitting	Hind end is in contact with the floor; front end is raised and supported by front legs.
Lying	Sow is lying (ventral or lateral). In ventral lying, the sow's belly region is in contact with the floor. In lateral lying, one side of the body is in contact with the floor and legs are extended to one side.

2.3.1. Skin Lesion Assessment

Lesion scores were used to evaluate levels of aggression in pigs, using methods adapted from Hodgkiss et al. [20]. For each treatment period, skin lesion scores were assessed on day 1 (before enrichment was provided) and on day 12. Lesion scores ranged from 0 (no injury) to 3 (severe injury), and were assessed on 11 regions (head, ear, neck, shoulder, top of back, tail, vulva, hind leg, side, udder and front leg) on both the right and left sides of the body (Figure 4). The scoring accounted for fresh injuries only, following the description; 0 = No injury (skin unmarked: no evidence of injury), 1 = Slight injury (<5 superficial wounds), 2 = Obvious injury (5–10 superficial wounds and/or <3 deep wounds), 3 = Severe injury (>10 superficial wounds, and/or >3 deep wounds).

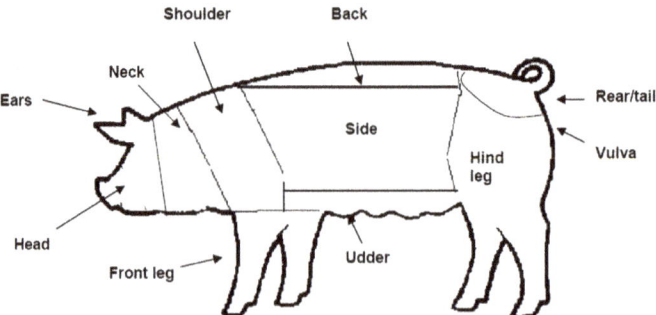

Figure 4. The animal's body was divided in to 11 areas as illustrated. Each area on both sides was assessed for skin lesions (Score = 0 to 3).

2.3.2. Cortisol in Saliva

Saliva samples were collected from the three Dom and three Sub sows per group at three time points: (1) The end of week 1 after application of first treatment; (2) day 11 of the second treatment (i.e., week 9 of gestation); and (3) on completion of the final treatment (week 14). Saliva samples were collected between 8 and 9 a.m. each day to control for diurnal variation in cortisol levels. Saliva was collected by allowing sows to chew on large cotton buds wrapped on a metal support until the bud was thoroughly moistened, about 30 to 60 s per sample. The moistened buds were placed in 15 mL centrifuge tubes (Fisher Scientific, Ottawa, ON, Canada) and centrifuged immediately for 15 min at $830 \times g$ (Beckman TJ-6 Centrifuge, Beckman Coulter, Mississauga, ON, Canada) to remove mucins and other particulate matter. Saliva samples were transferred to labeled storage tubes using disposable pipettes and stored at −20 °C until analysis.

The cortisol concentration in saliva was determined using the Salimetrics® Cortisol Enzyme Immunoassay Kits (Salimetrics, State College, PA, USA) for research in humans and animals. The Kit is a competitive immunoassay designed for quantitative measurement of salivary cortisol using a 96-well ELISA plate with spectrophotometric detection at 450 nm. The manufacturer's instructions were followed and each 96-well plate contained six standards, one zero and one nonspecific binding sample, two controls and 38 samples, all were run in duplicate. Intra-assay precision estimates on low and high standards (n = 20) gave values of 1.14 ± 0.05 and 0.16 ± 0.01 µg/dL (mean ± SD) and CV's of 4% and 5%, respectively. Inter-assay precision estimates on low and high standards (n = 20) gave values of 1.14 ± 0.05 and 0.18 ± 0.01 µg/dL (mean ± SD) and CV's of 4% and 9%, respectively.

2.4. Statistical Analysis

Measurements were done on 6 pens of sows, with four enrichment presentations (treatments) applied to each group in random sequence. Each treatment lasted 12 days, with sows' behaviour observed on days 1, 8, 10 and 12. Therefore, for all group behavior observations, n = 6 groups × 4 treatments × 4 observation days = 96 observations were analyzed using repeated measures models in SAS 9.3 (SAS Institute Inc., Cary, NC, USA).

Data manipulation: Pen was the experimental unit when analyzing outcome variables associated with enrichment use (location and postures) both for group level and focal sow photo scan observations. The proportion of time in each behaviour (defined in Table 1) was calculated as the proportions of scans (photos) where the behaviour was observed out of the total number of scans (photos) recorded during the 8 h observation period (48 scans per day). The number of sows performing each behaviour was calculated by the average proportion of sows that performed the behaviour when the behaviour was observed.

Sow was the experimental unit for lesion scores and cortisol analysis. For lesion scores, a combined skin lesion score was created by summing the lesion scores for all body regions. The effects of different enrichment presentations and treatment day (day 1 and day 11) were assessed using the total body score. To study lesion scores in different body regions, the 11 regions (Figure 4) were condensed in to four regions (head, shoulder, side and hind) and assessed for associations with independent variables enrichment treatment, social status and parity. Parity was categorised in three groups: Parity code 1 (parities 1 and 2), parity code 2 (parity 3) and parity code 3 (parities 4 to 7).

Data Analysis: Location and postural behavioural data were analysed using GLMM with beta regression models using SAS 9.3 to account for proportional data. Model fit was assessed by plotting the residuals. Fixed effects in the group behaviour model included enrichment treatment, day of treatment (day 1, 8, 10 and 12) and their interaction. The effects of social status on sow behaviour was studied in a similar model with social status added as a fixed effect and including two-way interactions of social status and treatment and social status and day of treatment. Sow group (replicate) and day of treatment were used as random effects to account for the correlation associated with repeated measurement of the same animals. Day of gestation was controlled for as each group (pen of sows) received all treatments, and the order of enrichment treatments was randomized. The significance level was set at $p < 0.05$.

Proc Mixed was used to assess the association between cortisol concentration and fixed factors including social status and parity category, and sow group (replicate) was the random effect. A similar model was used for lesion scores with measures for day 1 and 12 observations being analysed separately and individual sow ID was added as a repeated measure.

3. Results

3.1. Group Level Observations

3.1.1. Enrichment Use at Group Level

Significantly more sows were observed contacting enrichments in the Rotate and Stimulus treatments, compared to Constant treatment (Table 2). Day of observation also had significant effect on the number of sows contacting enrichment (Figure 5). On day 10, when straw was provided in the Rotate and Stimulus treatments, sows contacted enrichment more frequently than when rope (days 1 and 8) or wood (day 12) were provided ($n = 6$, F (3, 62), $p = 0.001$). Also, more sows contacted the straw than when wood (day 1, 8, 10 and 12) was provided in the Constant treatment (Figure 5). No significant interactions were found between enrichment treatment and day of observation.

Table 2. Effects of enrichment treatment and day of observation on time spent (proportion of observations) and number of sows (proportion of sows) in contact with enrichment, less than 1 m or greater than 1 m from enrichment.

Behaviour *	Treatment				p Value	
	Constant	Rotate	Stimulus	SEM	Treatment	Day **
Contacting enrichment:						
Time (prop. of obs.)	0.569	0.598	0.612	0.095	0.966	0.329
No. of sows (prop.)	0.112 [a]	0.200 [b]	0.182 [b]	0.019	0.041	0.001
Less than 1M:						
Time (prop. of obs.)	0.527	0.642	0.651	0.062	0.238	0.267
No. of sows (prop.)	0.123	0.168	0.159	0.028	0.371	0.388
Greater than 1M:						
Time (prop. of obs.)	0.999	0.988	0.988	0.001	0.358	0.299
No. of sows (prop.)	0.667 [a]	0.541 [b]	0.532 [b]	0.020	0.001	0.001

[ab] LSMeans with different a superscript within the same row are significantly different ($p < 0.05$). * Sows were observed during an 8 h period (08:00–16:00 h) on four days (1, 8, 10 and 12) using time lapse photos at 10 min intervals. ** Day = Day of observation (days 1, 8, 10 and 12).

3.1.2. Behaviour at Group Level

When sow postures (standing, sitting and lying) were studied at group level, sows were observed to spend more time sitting in the Control (no enrichments), Constant and Rotate treatments compared to the Stimulus treatment ($n = 6$, F (3, 89) = 2.99, $p = 0.035$, Table 3). There was a tendency for more sows to stand in Rotate and Stimulus than in Control or Constant treatments ($p = 0.081$, Table 3).

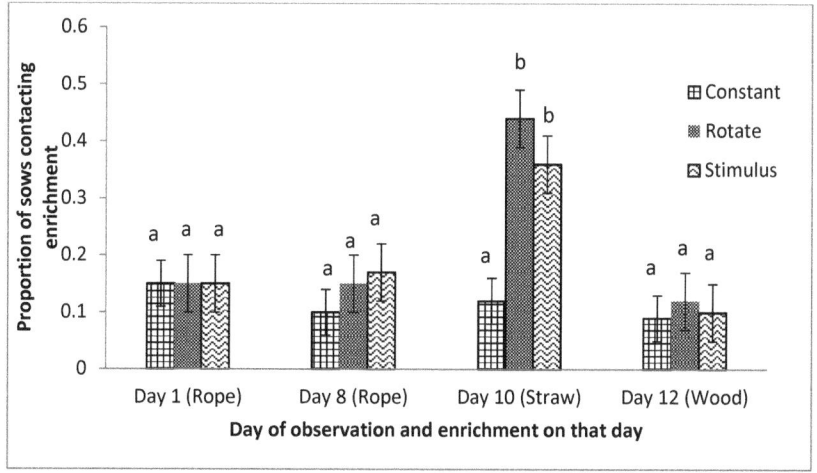

Figure 5. Proportion of sows contacting enrichments (rope, wood or straw) and day of observation (days 1, 8, 10 and 12) in the Constant, Rotate and Stimulus treatments. Note: sows in the Constant treatment received the wood enrichment on all days. Sows were observed during an 8 hour period (08:00–16:00 h) on four days (1, 8, 10 and 12) using time lapse photos at 10 min intervals. Bars without a common superscript are significantly different ($p < 0.05$). [a,b] LSMeans with a different superscript are significantly different ($p < 0.05$).

Table 3. Effects of enrichment treatment and day of observation on the postures of sows observed in photo scans *. LS Means of duration (proportion of observations) and number of sows in each posture (proportion of sows).

Behaviour *	Treatment				SEM	p-Value	
	Control	Constant	Rotate	Stimulus		Treatment	Day **
Standing:							
Time (prop. of obs.)	0.965	0.944	0.938	0.932	0.022	0.936	0.896
No. of sows (prop.)	0.444	0.466	0.513	0.513	0.022	0.081	0.061
Sitting:							
Time (prop. of obs.)	0.283 [a]	0.307 [a]	0.219 [a]	0.181 [b]	0.041	0.035	0.201
No. of sows (prop.)	0.057	0.069	0.058	0.061	0.009	0.352	0.547
Lying:							
Time (prop. of obs.)	0.954	0.970	0.940	0.939	0.012	0.193	0.027
No. of sows (prop.)	0.343	0.314	0.298	0.294	0.023	0.430	0.492

* Sows were observed during an 8 h period (08:00–16:00 h) on four days (1, 8, 10 and 12) using time lapse photos at 10 min intervals. ** Day = day of observation (days 1, 8, 10 and 12). [a,b] LSMeans with a different superscript within the same row are significantly different ($p < 0.05$).

3.2. Focal Pig Observations

The parity distribution of Dom and Sub focal sows was: Dominants- parity 2: 39%; 3: 11%, 4: 17%, 5: 17%, 6: 11% and 7: 5% and Subordinates- Parity 1: 11%, 2: 28%, 3: 22%, 4: 22% and 5: 16% (18 Dom and 18 Sub sows in total). Parity had no effect on any measures in focal sows.

3.2.1. Enrichment Use by Focal Sows

The effect of social status was studied by observing a sample of six focal sows (3 Dom and 3 Sub) per group of 20 ± 2 sows. Sub sows spent more time close to enrichments (<1 m) than Dom ($p = 0.001$), and also more time >1 m from enrichments than Dom sows ($p = 0.001$, Table 4).

Table 4. Effects of social status and day of observation on time spent (proportion of observations) and number of sows (proportion of sows) in contact with enrichment, less than 1 m or greater than 1 m from enrichment.

Behaviour *	Social Status **			p-Value ***	
	Dom	Sub	SEM	SS	Day
Enrichment contact					
Time (prop. of obs.)	0.279	0.314	0.024	0.172	0.001
No. of sows (prop.)	0.386	0.397	0.009	0.939	0.001
Less than 1 m					
Time (prop. of obs.)	0.137 [a]	0.203 [b]	0.015	0.001	0.001
No. of sows (prop.)	0.371	0.371	0.009	0.992	0.001
Greater than 1 m					
Time (prop. of obs.)	0.854 [a]	0.905 [b]	0.016	0.001	<0.001
No. of sows (prop.)	0.563 [a]	0.591 [b]	0.013	0.042	0.001

[ab] LS Means with a different superscript within the same row are significantly different ($p < 0.05$). * Sows were observed during an 8 h period (08:00–16:00 h) on four days (1, 8, 10 and 12) using time lapse photos at 10 min intervals. ** Social Status LS Means: Dom = Dominant; Sub = Subordinate. *** SS = Social Status, Day = Day of observation (1, 8, 10 and 12).

3.2.2. Behaviour of Focal Sows

Social status had a significant effect on standing behaviour, with Sub sows spending more time standing and higher number of sows performing this activity when observed (Table 5). No interactions were found among social status, treatment or day.

Table 5. Effects of social status enrichment treatment and day on focal sow postures observed in photo scans. LS Means of duration (proportion of observations) and number of sows (proportion of sows) in each posture.

Behaviour *	Social Status **			p-Value		
	Dom	Sub	SEM	SS	Treat.	Day
Standing						
Time (prop. of obs.)	0.771 [a]	0.824 [b]	0.022	0.002	0.178	0.728
No. of sows (prop.)	0.561 [a]	0.585 [b]	0.013	0.007	0.023	0.407
Sitting						
Time (prop. of obs.)	0.055	0.059	0.004	0.619	0.299	0.977
No. of sows (prop.)	0.342	0.334	0.041	0.233	0.674	0.348
Lying						
Time (prop. of obs.)	0.489	0.512	0.040	0.409	0.003	0.522
No. of sows (prop.)	0.447	0.437	0.015	0.440	0.002	0.899

[ab] LS Means with a different superscript within the same row are significantly different ($p < 0.05$). * Sows were observed during an 8 h period (08:00–16:00 h) on four days (1, 8, 10 and 12) using time lapse photos of 10 min intervals. ** Social Status LS Means: Dom = Dominant; Sub = Subordinate.

3.2.3. Skin Lesions

Sub sows received more lesions on the shoulder area than Dom sows ($p < 0.001$, Table 6), and had numerically higher lesion scores on all body regions. Social status had no significant effect on head, side or hind lesions (data not shown). There were significant differences in lesions scores among treatments (Table 7). On treatment day 1, sows in the Stimulus treatment had significantly higher lesion scores than sows in all other treatments. However, because day 1 lesions were assessed as a baseline, before enrichment was given, these differences cannot be attributed to the enrichment treatment. Sows may have fought more during the two-day interval between treatments when no enrichment was provided. The absence of significant treatment effects on day 12 lesion scores is arguably a more important result. No interactions were found between social status and treatment for lesion scores.

Table 6. Effects of social status on skin lesion scores in focal sows. Sows measured on treatment days 1 and 12.

	Social Status *			
Lesion Scores **	Dom	Sub	SEM	p-Value
Day 1 Total	2.74 [a]	4.46 [b]	0.44	0.004
Shoulder Day 1	0.66 [a]	1.17 [b]	0.12	0.001
Day 12 Total	3.61	4.24	0.36	0.293
Shoulder Day 12	0.72	0.92	0.12	0.214

* Social Status LS Means: Dom = Dominant; Sub = Subordinate. ** Lesion scores were done using a scale of zero to three (0 = no injury, and 3 = severe injury) on 11 regions on the right and left side of sows. Total = sum of lesions for all 22 body regions. [ab] LS Means with a different superscript within the same row are significantly different ($p < 0.05$).

Table 7. Effects of enrichment treatments on skin lesion scores in focal sows. Lesion scores were measured on day 1 (before enrichment) and day 12 of each treatment period.

Lesion Scores *	Treatments					
	Control	Constant	Rotate	Stimulus	SEM	p-Value
Day 1 Total	3.54 [a]	2.50 [a]	3.32 [a]	5.41 [b]	0.57	0.002
Shoulder Day 1	1.11 [a]	0.47 [ab]	0.83 [ab]	1.40 [c]	0.18	0.001
Side Day 1	0.63 [a]	0.33 [a]	0.58 [a]	0.91 [b]	0.13	0.028
Day 12 Total	4.01	3.42	4.50	4.11	0.27	0.446
Shoulder Day 12	0.77	0.63	1.00	0.80	0.16	0.465
Side Day 12	0.40	0.46	0.51	0.41	0.11	0.908

[ab] LS Means with a different superscript within the same row are significantly different ($p < 0.05$). * Lesion scores were done using a scale of zero to three (0 = no injury, and 3 = severe injury) on 11 regions on the right and left side of sows. Total = sum of lesions for all 22 body regions.

3.2.4. Cortisol in Focal Sows

Social status had a significant effect on the cortisol levels of sows, with Sub sows having significantly higher cortisol concentrations than Dom sows (Table 8). Sampling time point also had a significant effect, with cortisol levels in late gestation being significantly higher than in early or mid-gestation.

Table 8. Effect of social status and sample time point on salivary cortisol levels (LS Means, μg/dL) in focal sows.

Factor	Cortisol Concentration	SEM	p-Value
Social Status:			
Dominant	0.25 [a]	0.04	0.002
Subordinate	0.43 [b]	0.04	
Sample time point:			
Early gestation (week 5)	0.22 [a]	0.04	0.005
Mid gestation (week 9)	0.39 [a]	0.05	
Late gestation (week 14)	0.43 [b]	0.04	

[ab] LS Means with different superscripts within a factor are significantly different ($p < 0.05$).

4. Discussion

This study compared the effects of different types of enrichments and presentation methods on sow behavior, and the influence of social status on enrichment interaction and behaviour. Two enrichment presentation methods (Rotate and Stimulus) were used to compare the effects of different enrichment materials (straw, rope and wood) on the time spent interacting with enrichment and sows' proximity to enrichment. The results indicate that sows interacted with straw, rope and wood in decreasing order. This is in agreement with previous studies which found that straw produced a greater behavioural response compared to object enrichments [10,21,22]. Elmore et al. [23] observed that sows spent more time using a straw rack compared to mushroom compost, suggesting a preference for consumable enrichments over non-consumable enrichments. Straw was included in the present study as a "positive control" for comparison with object enrichments. As predicted, more sows were observed contacting enrichment when straw was provided. The greater use of straw enrichment in the current study was also influenced by the fact that straw was provided on the floor, in a more diffuse manner than hanging objects, allowing more sows to interact with the material at once. This study also showed that providing 300 g of straw per sow in partially slatted pens did not cause any blockage of the liquid manure system. Most of the material was consumed by sows and only a small amount entered the pits. Limitations to the study design include the fact that the rotation of rope, straw and wood was always done in the same order, which could have influenced the pigs' behavior, and that the two pen designs used may have produced different effects but were not included in the model due to low sample size.

Although rope and wood enrichments elicited less contact by sows than straw in this study, their usefulness as enrichment materials cannot be discounted. Previous studies have also reported 'hanging rope' as an effective enrichment strategy [10,14,17]. Multiple studies indicate that for pigs to show sustained interest in a particular enrichment, the enrichment should be ingestible, flexible and destructible [8,13,18]. Rope and wood satisfy at least two of the described characteristics (flexible and destructible) and hence can be good objects for provision of enrichment. However, the accessibility of enrichment material (ratio of enrichments to sows) is another factor that must be considered, as limited access to desirable forms of enrichment may increase competition and aggression.

Pigs' interest in enrichment can be sustained by changing objects or renewing substrates on a regular basis [8,24] thereby mitigating effects of habituation. We hypothesised that frequent rotation of the enrichments (as in the Rotate and Stimulus treatments) would maintain novelty (reduce habituation) and increase the activity level of sows. The results show that rotation of enrichment objects resulted in significantly more sows within 1M and contacting enrichments. It is likely that changing enrichments every three to four days was responsible for this result. In a study of growing pigs, Van de Perre et al. [8] tested a sequence of enrichment objects by rotating seven different enrichment objects every week throughout the growing period and found a significant reduction in pen mate biting behaviour and wounds compared to presentation of a chain enrichment only. Pigs' interaction with enrichment was greatest on day 1, and declined significantly by day 5 of each rotation [8]. Grifford et al. [25] studied the ability of pigs to recognize objects following different exposure periods (10 min vs. 2 days) and delays before re-exposure. Based on the results they suggested restricting object exposure to less than two days in rotated objects, and delaying re-exposure to the same object by more than one week to enhance the exploratory value of enrichment. Since both of these studies used growing pigs, generalizing the results to sow behaviour may not be appropriate. Since novelty played a role in increasing the interest of sows in this study, further studies should examine how different rotation frequencies and delays to re-exposure of a particular enrichment can influence sow behaviour, as well as the optimal ratio of enrichments to sows.

Pairing enrichment with another stimulus has been used to increase the value of enrichments by some researchers. For example, the study by Dudink et al. [9] used an associative stimulus in conjunction with enrichment provision, which resulted in anticipatory excitement and increased play behaviour in young pigs. By presenting enrichments in conjunction with an associative sound stimulus in the current study, we hypothesized that the stimulus would increase the initial response of sows to the delivery of new enrichments. However, no evidence of an increased initial response was found. Douglas et al. [26], paired enrichments with an auditory stimulus in gilts and measured an improvement in their cognitive performance. In both the Dudink et al. [9] and Douglas et al. [26] studies, pigs were given straw in addition to food rewards in the form of mixed seeds or sliced apple, respectively, after the sound cues. The studies also used an intensive training period for pigs to learn the association between the rewards and sound cues. There was 30 s delay after the sound cue before rewards were given and the sequence of the rewards and auditory cues was randomised. Our goal in this study was to simulate enrichment provision as it would be implemented in commercial practice, so no training phase was used. Because we did not use a training phase, it is unclear whether sows made a clear association between the sound cue and enrichment provision. Future studies should include a training phase so that the association is learned, and should also recognise the potential for an associative stimulus to increase competition and aggression if access to enrichment is limited.

One of the goals of providing enrichment is to increase the frequency and duration of normal behaviours, such as increasing activity as represented by standing behaviour [27–29]. If enrichment treatments are effective, we would expect to see a higher level of activity when enrichment is presented than in a barren pen environment. In terms of sows' postures, rotating enrichments (Rotate and Stimulus treatments) resulted in more sows standing, and the Stimulus treatment reduced the duration of sitting, suggesting that sow activity increased when enrichments were rotated. Gestating sows are generally inactive, spending roughly 80% of their time lying [30]. However, a moderate level of activity should improve fitness and be beneficial, as shown in studies comparing stall and group housed sows [31]. Further research should explore potential benefits of increased activity and "appropriate" levels of activity for gestating sows.

Regarding the effects of social status on enrichment interaction, we hypothesised that if enrichment was considered a valuable resource, dominant sows would have greater access to enrichment or access it at preferred times, compared to subordinates. However, the results show that, sows gained access to enrichments regardless of social status, and in fact Sub sows spent more time near enrichments than Dom sows. One explanation for this may be that Dom sows may have been more preoccupied with guarding the ESF feeder, as the feeder is clearly an important resource within the pen. Furthermore, because social status was determined based on a feed competition test, it may be that differences in social status are confounded with feeding motivation. Elmore et al. [17] found no significant effect of social status on stall-housed gestating sows' motivation to access an enriched group pen. In their study, both dominant and subordinate sows showed similar levels of an operant response and similar latency to press a panel to access enrichment objects. However, once the sows were released into a group setting, social status had a significant effect on enrichment use, with dominant sows spending more time interacting with the objects than subordinates.

Elmore et al. [17] observed significant effects of social status on active behaviour such as standing and inactive behaviours, with dominant sows being more active and standing more compared to subordinate sows. However, in this study, subordinate sows spent more time standing than dominants. Because behaviour observations were collected using a camera focused on the enrichments, the behaviour of Dom sows in other pen areas was not recorded. As suggested previously, Dom sows appear to have been more focused on the feeder area, which may explain the greater enrichment use and increased activity level of subordinate sows in this study. It also suggests that either there was sufficient availability of enrichment objects (three per 20 sows) to reduce competition, or that dominant sows did not view the enrichments as highly valuable.

Lesion scores have been used in previous studies to determine the aggression level of pigs [20]. Hodgkiss et al. [20] and Stukenborg et al. [32] demonstrated that the severity, duration and frequency of overt aggression or fighting can be estimated by looking at the number of scratches or injuries on pigs, especially injuries to the anterior region of their body. However, aggression is also influenced by factors such as the familiarity of pigs, space allowance, group size and composition, pen design, time of day, food and bedding [33]. According to these authors, aggression or its absence in the social environment is affected largely by the degree of competition over resources. For example, an increase in aggression was found when group-housed grower-finisher pigs were given limited access to a highly valued enrichment—straw [34]. We hypothesised that subordinate sows would receive more skin lesions due to aggression than dominant sows. Since the lesion results are consistent with that hypothesis, we conclude that subordinate sows experienced more aggression than dominants in this management system, however, it should be noted that the scores were consistently low.

Subordinate sows in this study had significantly higher salivary cortisol concentrations compared to dominant sows. Because all sows received all enrichment treatments (in random order) and saliva samples were only collected at three time points, it was impossible to determine effects of the enrichment treatments on cortisol. Usually the social hierarchy is established in gestation sows within 24–48 h after mixing in a group housing type of management. Studies have found an increase in cortisol concentration shortly after mixing in group housing due to social aggression, and after a period of 48 h cortisol concentration returns to baseline [35]. There is evidence to indicate that if there is no competition for resources, then there should be no difference in cortisol concentration between subordinate and dominant sows after a week. O'Connell et al. [36] examined the effects of social status on the welfare of sows in dynamic groups. They found no difference in salivary cortisol concentration among subordinate and dominant sows at one week after mixing. However, if there are insufficient resources such as space, food or environmental enrichment for the group, then stress levels remain elevated [37] particularly in subordinate sows. This in combination with lesion results indicates that subordinate sows experienced greater stress compared to dominants in this management system. Another possible explanation may be that due to ESF feeding (here with a 9 a.m. reset time), the diurnal cortisol rhythm of subordinate sows may be shifted. Further research to determine whether the ESF system affects the circadian rhythm of dominant and subordinate sows differently would be of interest. All sows showed higher salivary cortisol concentrations at 14 weeks gestation than at five or nine weeks. The increase in cortisol levels during late gestation was expected as this is known to occur as sows (and other species) approach parturition [38–40].

5. Conclusions

The different enrichment materials (rope, wood and straw) and how they were presented to sows both had significant effects on the total number of sows contacting enrichment. Straw enrichment provided on the pen floor produced the greatest response to object enrichments, however, its presence only in the Rotate and Stimulus treatments represents a confounding effect in this study. Sows also made use of the rope and wood on chains enrichments, with wood being the least contacted. Regularly changing enrichment materials (as in Rotation and Stimulus treatments) increased sows' response to enrichment. These findings demonstrate that novelty and the type of material provided play an important role in increasing attractiveness and sustaining sows interest in enrichment. Future research should explore the most effective rotation schedule and the optimal ratio of enrichments to sows.

Social hierarchy (dominant or subordinate) had a significant effect on enrichment use, with subordinate sows having greater interaction with enrichment. We suggest that this may be due to dominant sows valuing the ESF feeder more than enrichments. Subordinate sows in this study had increased skin lesions, as well as higher salivary cortisol concentrations suggesting that these individuals experienced greater stress than dominants in this management system. Longer duration studies are needed to better evaluate the interaction between social status and enrichment treatments.

Author Contributions: Conceptualization, J.A.B., Y.M.S. and L.M.C.; methodology, V.K., J.A.B., Y.M.S. and L.M.C.; software, L.L. and V.K.; validation, J.A.B., L.L. and C.R.; formal analysis, C.R., J.A.B. and Y.M.S.; investigation, L.L. and L.M.C.; resources, L.M.C.; data curation, L.L. and C.R.; writing—original draft preparation, C.R.; writing—review and editing, J.A.B.; visualization, C.R. and J.A.B.; supervision, J.A.B. and L.M.C.; project administration, L.M.C.; funding acquisition, L.M.C.

Funding: This research was funded by Swine Innovation Porc, grant number 1231, as part of Agriculture and Agri-food Canada's 'Growing Forward 2' Program.

Acknowledgments: The research team wishes to acknowledge the animal care and cooperation of farm staff at the University of Manitoba's Glen Lea Research Centre.

Conflicts of Interest: The authors declare no conflict of interest. The funders had no role in the design of the study; in the collection, analyses, or interpretation of data; in the writing of the manuscript, or in the decision to publish the results.

References

1. Van de Weerd, H.A.; Day, J.E.L. A review of environmental enrichment for pigs housed in intensive housing systems. *Appl. Anim. Behav. Sci.* **2009**, *116*, 1–20. [CrossRef]
2. Karlen, G.A.M.; Hemsworth, P.H.; Gonyou, H.W.; Fabrega, E.; Strom, A.D.; Smits, R.J. The welfare of gestating sows in conventional stalls and large groups on deep litter. *Appl. Anim. Behav. Sci.* **2007**, *105*, 87–101. [CrossRef]
3. Whittaker, X.; Edwards, S.A.; Spoolder, H.A.M.; Lawrence, A.B.; Corning, S. Effects of straw bedding and high fibre diets on the behaviour of floor fed group-housed sows. *Appl. Anim. Behav. Sci.* **1999**, *63*, 25–39. [CrossRef]
4. Maes, D.; Pluym, L.; Peltoniemi, O. Impact of group housing of pregnant sows on health. *Porc. Health Manag.* **2016**, *2*, 17. [CrossRef]
5. Chapinal, N.; de la Torre, J.L.R.; Cerisuelo, A.; Gasa, J.; Baucells, M.D.; Coma, J.; Vidal, A.; Manteca, X. Evaluation of welfare and productivity in pregnant sows kept in stalls or in 2 different group housing systems. *J. Vet. Behav.* **2010**, *5*, 82–93. [CrossRef]
6. Zhou, Q.; Sun, Q.; Wang, G.; Zhou, B.; Lu, M.; Marchant-Forde, J.; Yong, J.N.; Zhao, R. Group housing during gestation affects the behaviour of sows and the physiological indices of offspring at weaning. *Animal* **2014**, *8*, 1162–1169. [CrossRef] [PubMed]
7. Day, J.E.L.; Van de Weerd, H.A.; Edwards, S.A. The effect of varying lengths of straw bedding on the behaviour of growing pigs. *Appl. Anim. Behav. Sci.* **2008**, *109*, 249–260. [CrossRef]
8. Van de Perre, V.; Driessen, B.; Thielen Van, J.; Verbeke, G.; Geers, R. Comparison of pig behaviour when given a sequence of enrichment objects or a chain continuously. *Anim. Welf.* **2011**, *20*, 641–649.
9. Dudnik, S.; Simonse, H.; Marks, I.; de Jonge, F.H.; Spruijt, B.M. Announcing the arrival of enrichment increases play behaviour and reduces weaning-stress-induced behaviours of piglets directly after weaning. *Appl. Anim. Behav. Sci.* **2006**, *101*, 86–101. [CrossRef]
10. Van de Weerd, H.A.; Docking, C.M.; Day, J.E.L.; Breuer, L.K.; Edwards, S.A. Effects of species-relevant environmental enrichment on the behaviour and productivity of finishing pigs. *Appl. Anim. Behav. Sci.* **2006**, *99*, 230–247. [CrossRef]
11. Lyons, C.A.P.; Bruce, J.; Fowlerb, V.R.; English, P.R. A comparison of productivity and welfare of growing pigs' intensive systems. *Livest. Prod. Sci.* **1995**, *43*, 265–274. [CrossRef]
12. Grimberg-Henrici, C.G.E.; Vermaak, P.; Bolhuis, J.E.; Nordquist, R.E.; van der Staay, F.J. Effects of environmental enrichment on cognitive performance of pigs in a spatial holeboard discrimination task. *Anim. Cogn.* **2016**, *19*, 271–283. [CrossRef] [PubMed]
13. Courboulay, V. Enrichment materials for fattening pigs: Summary of IFIP trials. *Cahiers de l'IFIP* **2014**, *1*, 47–56.
14. Horback, K.M.; Pierdon, M.K.; Parsons, T.D. Behavioural preference for different enrichment objects in a commercial sow herd. *Appl. Anim. Behav. Sci.* **2016**, *184*, 7–15. [CrossRef]
15. Petherick, J.C.; Blackshaw, J.K. A review of the factors influencing the aggressive and agonistic behaviour of domestic pig. *Aust. J. Exp. Agric.* **1987**, *27*, 605–611. [CrossRef]
16. Trickett, S.L.; Jonathan, H.G.; Edwards, S.A. The role of novelty in environmental enrichment for the weaned pig. *Appl. Anim. Behav. Sci.* **2009**, *116*, 45–51. [CrossRef]
17. Elmore, M.R.P.; Garner, J.P.; Johnson, A.K.; Kirkden, R.D.; Richert, B.T.; Pajor, E.A. Getting around social status: Motivation and enrichment use of dominant and subordinate sows in a group setting. *Appl. Anim. Behav. Sci.* **2011**, *133*, 154–163. [CrossRef]
18. Van de Weerd, H.A.; Docking, C.M.; Day, J.E.L.; Avery, P.J.; Edwards, S.A. A systematic approach towards developing environmental enrichment for pigs. *Appl. Anim. Behav. Sci.* **2003**, *84*, 101–118. [CrossRef]

19. Anderson, I.L.; Boe, K.E.; Kristiansen, A.L. The influence of different feeding arrangements and food type on competition at feeding in pregnant sows. *Appl. Anim. Behav. Sci.* **1999**, *65*, 91–104. [CrossRef]
20. Hodgkiss, N.J.; Eddison, J.C.; Brooks, P.H.; Bugg, P. Assessment of the injuries sustained by pregnant sows housed in groups using electronic feeders. *Vet. Rec.* **1998**, *143*, 604–607. [CrossRef]
21. Scott, K.; Taylor, L.; Gill, B.P.; Edwards, S.A. Infuence of different types of environmental enrichment on the behaviour of fnishing pigs in two different housing systems: Hanging toy versus rootable substrate. *Appl. Anim. Behav. Sci.* **2006**, *99*, 222–229. [CrossRef]
22. Elmore, M.R.P.; Garner, J.P.; Johnson, A.K.; Kirkden, R.D.; Richert, B.T.; Patterson-kane, E.G.; Pajor, E.A. Differing results for motivation tests and measures of resource use: The value of environmental enrichment to gestating sows housed in stalls. *Appl. Anim. Behav. Sci.* **2012**, *141*, 9–19. [CrossRef]
23. Elmore, M.R.P.; Garner, J.P.; Johnson, A.K.; Kirkden, R.D.; Richert, B.T.; Pajor, E.A. If you knew what was good for you! The value of environmental enrichments with known welfare benefits is not demonstrated by sows using operant techniques. *J. Appl. Anim. Welf. Sci.* **2012**, *15*, 254–271. [CrossRef] [PubMed]
24. Grandin, T.; Curtis, S.E.; Greenough, W.T. Effects of rearing environment on the behaviour of young pigs. *J. Anim. Sci.* **1983**, *57*, 137.
25. Gifford, A.K.; Cloutier, S.; Newberry, R.C. Objects as enrichment: Effects of object exposure time and delay interval on object recognition memory of the domestic pig. *Appl. Anim. Behav. Sci.* **2007**, *107*, 206–217. [CrossRef]
26. Douglas, C.; Bateson, M.; Walsh, C.; Anais, B.; Edwards, S.A. Environmental enrichment induces optimistic cognitive biases in pigs. *Appl. Anim. Behav. Sci.* **2012**, *139*, 65–73. [CrossRef]
27. Novak, M.A.; Kinsey, J.H.; Jorgensen, M.J.; Hazen, T.J. Effects of puzzle feeders on pathological behavior in individually housed rhesus monkeys. *Am. J. Primatol.* **1998**, *46*, 213–227. [CrossRef]
28. Beattie, V.E.; Walker, N.; Sneddon, I.A. Effect of rearing environment and change of environment on the behaviour of gilts. *Appl. Anim. Behav. Sci.* **1995**, *46*, 57–65. [CrossRef]
29. Scott, K.; Chennells, D.J.; Armstrong, D.; Taylor, L.; Gill, B.P.; Edwards, S.A. The welfare of finishing pigs under different housing and feeding systems: Liquid versus dry feeding in fully-slatted and straw-based housing. *Anim. Welf.* **2007**, *16*, 53–62.
30. Tuyttens, F.A.M.; Wouters, F.; Stuelens, E.; Sonck, B.; Duchateau, L. Synthetic lying mats may improve comfort of gestating sows. *Appl. Anim. Behav. Sci.* **2008**, *114*, 76–85. [CrossRef]
31. Broom, D.M.; Mendl, M.T.; Zanella, A.J. A comparison of the welfare of sows in different housing conditions. *Anim. Sci.* **1995**, *61*, 369–385. [CrossRef]
32. Stukenborg, A.; Traulsen, I.; Stamer, E.; Puppe, B.; Krieter, J. The use of a lesion score as an indicator for agonistic behaviour in pigs. *Arch. Anim. Breed.* **2012**, *55*, 163–170. [CrossRef]
33. Arey, D.S.; Edwards, S.A. Factors influencing aggression between sows after mixing and the consequence for welfare and production. *Livest. Prod. Sci.* **1998**, *56*, 61–70. [CrossRef]
34. Olsen, W.A.; Simonsen, H.B.; Dybkjær, L. Effect of access to roughage and shelter on selected behavioural indicators of welfare in pigs housed in a complex environment. *Anim. Welf.* **2002**, *11*, 75–87.
35. Pedersen, L.J.; Rojkittikhun, T.; Einarsson, S.; Edqvist, L.E. Post weaning grouped sows: Effects of aggression on hormonal patterns and oestrous behaviour. *Appl. Anim. Behav. Sci.* **1993**, *36*, 25–39. [CrossRef]
36. O'Connell, N.E.; Beattie, V.E.; Moss, B.W. Influence of social status on the welfare of sows in static and dynamic groups. *Anim. Welf.* **2003**, *12*, 239–249.
37. Cornale, P.; Macchi, E.; Miretti, S.; Renna, M.; Lussiana, C.; Perona, G.; Mimosi, A. Effects of stocking density and environmental enrichment on behavior and fecal corticosteroid levels of pigs under commercial farm conditions. *J. Vet. Behav.* **2015**, *10*, 569–576. [CrossRef]
38. Bronwyn, S.; Karlen, M.G.; Morrison, R.; Gonyou, H.W.; Butler, K.L.; Kerswell, K.J.; Hemsworth, P.H. Effects of stage of gestation at mixing on aggression, injuries and stress in sows. *Appl. Anim. Behav. Sci.* **2015**, *165*, 40–46.

39. Barnett, J.L.; Winfield, C.G.; Cronin, G.M.; Hemsworth, P.H.; Dewar, A.M. The effect of individual and group housing on behavioral and physiological responses related to the welfare of pregnant pigs. *Appl. Anim. Behav. Sci.* **1985**, *14*, 149–161. [CrossRef]
40. Hay, M.; Meunier-Salaün, M.C.; Brulaud, F.; Monnier, M.; Mormède, P. Assessment of hypothalamic pituitary-adrenal axis and sympathetic nervous system activity in pregnant sows through the measurement of glucocorticoids and catecholamines in urine. *J. Anim. Sci.* **2000**, *78*, 420–428. [CrossRef] [PubMed]

© 2019 by the authors. Licensee MDPI, Basel, Switzerland. This article is an open access article distributed under the terms and conditions of the Creative Commons Attribution (CC BY) license (http://creativecommons.org/licenses/by/4.0/).

Article

The Effect of Straw, Rope, and Bite-Rite Treatment in Weaner Pens with a Tail Biting Outbreak

Helle Pelant Lahrmann [1],*, Julie Fabricius Faustrup [2], Christian Fink Hansen [1], Rick B. D'Eath [3], Jens Peter Nielsen [2] and Björn Forkman [2]

1 SEGES, Danish Pig Research Centre, Agro Food Park 15, 8200 Aarhus, Denmark; CFHA@SEGES.dk
2 Department of Veterinary and Animal Sciences, University of Copenhagen, Grønnegårdsvej 8, 1870 Frederiksberg, Denmark; jffaustrup@gmail.com (J.F.F.); jpni@sund.ku.dk (J.P.N.); bjf@sund.ku.dk (B.F.)
3 SRUC, West Mains Road, Edinburgh EH9 3JG, UK; rick.death@sruc.ac.uk
* Correspondence: HLA@SEGES.dk; Tel.: +45-3339-4933

Received: 13 May 2019; Accepted: 13 June 2019; Published: 17 June 2019

Simple Summary: Young pigs can bite each other's tails, which is a welfare problem. It begins suddenly and spreads like an "outbreak". Pig farmers use various methods to prevent tail biting, but if prevention fails, a cure is needed, and there has been little scientific research into how best to stop an outbreak. In a study with 65 groups of young pigs, we tested three methods of stopping tail biting outbreaks which could be practical to use on commercial farms: (1) straw (small amount on the floor), (2) rope, and (3) Bite-Rite (a hanging plastic device with chewable rods). All provided some distraction, but straw stopped an increase in tail injuries more often (75%) than the Bite-Rite (35%), with rope intermediate (65%). Watching the pigs' behaviour showed that they preferred to interact with rope than the Bite-Rite. We also saw that interacting with other pigs' tails increased after a week with the Bite-Rite but not with rope or straw. Overall straw worked best, but future studies may find even more effective ways to stop tail biting outbreaks, once they begin.

Abstract: Tail biting in pigs is an injurious behaviour that spreads rapidly in a group. We investigated three different treatments to stop ongoing tail biting outbreaks in 65 pens of 6–30 kg undocked pigs (30 pigs per pen; SD = 2): (1) straw (7 g/pig/day on the floor), (2) rope, and (3) Bite-Rite (a hanging plastic device with chewable rods). Pigs were tail scored three times weekly, until an outbreak occurred (four pigs with a tail wound; day 0) and subsequently once weekly. After an outbreak had occurred, a subsequent escalation in tail damage was defined if four pigs with a fresh tail wound were identified or if a biter had to be removed. Straw prevented an escalation better (75%) than Bite-Rite (35%; $p < 0.05$), and rope was intermediate (65%). Upon introduction of treatments (day 0), pigs interacted less with tails than before (day −1; $p < 0.05$). Behavioural observations showed that pigs engaged more with rope than Bite-Rite ($p < 0.05$). Bite-Rite pigs (but not straw or rope) increased their interaction with tails between day 0 and day 7 ($p < 0.05$). Straw was the most effective treatment. However, further investigations may identify materials or allocation strategies which are more effective still.

Keywords: pigs; swine; weaners; behaviour; tail injury; tail biting outbreak; enrichment material; straw; rope; Bite-Rite

1. Introduction

Tail biting in pigs is an abnormal behaviour and has been reported both in conventional and in free-range/organic productions [1,2]. Tail biting by single pigs may, if not identified in the early stages, develop into a tail biting outbreak [3]. European Food Safety Authority (EFSA) [4] defined a tail biting outbreak as occurring when the tail biting intensifies, leading to several tail damaged pigs and the tail biting will continue, if an intervention is not conducted. If the tail biting continues within a group of pigs, the biting behaviour may lead to severe injuries with tail loss and infections followed by carcass condemnation at the abattoir [5].

It is well established that giving pigs access to enrichment reduces tail damage [6–9]. However, although there has been considerable research into risk factors and preventive treatments, only one experimental study, Zonderland et al. [8], has investigated the curative effect of interventions in pens with a tail biting outbreak. Systematic studies evaluating the effect of different curative treatments in pens with a tail biting outbreak are therefore needed [3,9]. Curative treatments in this context refer to interventions aiming at stopping or reducing tail biting behaviour and tail damage in pens after a tail biting outbreak has already started. The most frequently studied preventive enrichment material is straw on the floor in various amounts (reviewed by D'Eath et al. [9], Brunberg et al. [10]). To avoid the problems of loose straw in fully or part-slatted systems as either a preventive or curative treatment, hanging materials could be used. However, the effect of hanging materials on tail damage during an outbreak has not been investigated in previous studies (reviewed by D'Eath et al. [9]). D'Eath et al. [9] discussed that in order to avoid waste, as well as problems with straw blocking the slats and slurry system, litter material such as straw is only practicable in pens with a solid or part-solid floor and in small amounts. Other solutions are therefore needed for production systems without a solid floor to stop tail biting, once an outbreak has begun.

Scientific knowledge of the efficacy of different intervention strategies as curative measures for tail damage during a tail biting outbreak is crucial to reduce the negative welfare impact of the outbreak and to support evidence-based advice for farmers. The aim of the present study was therefore to examine the effect of either a small amount of straw on the floor, rope, or Bite-Rite (a hanging plastic device with chewable rods) on tail damage and behaviour in pens with a tail biting outbreak. These enrichment types were chosen due to their possible practical implementation, if the material successfully ceased tail damage. As previous studies reported, less use of enrichment in pens with Bite-Rite compared to straw [11] and because rope is more destructible, it was hypothesised that straw on the floor and rope would reduce tail damage more efficiently than Bite-Rite.

2. Materials and Methods

This experiment was a continuation of the work presented in Lahrmann et al. [12] using the same study subjects. However, while Lahrmann et al. [12] dealt with identifying behavioural changes before tail biting outbreaks, the current study focused on the effect of interventions in pens with tail biting outbreaks. The study was carried out at a commercial piggery from November 2015 to February 2016.

2.1. Ethical Consideration

This experiment was conducted in accordance with the guidelines of the Danish Ministry of Justice, Act No. 382 (June 10, 1987) and Acts 333 (May 19, 1990), 726 (September 9, 1993) and 1016 (December 12, 2001) with respect to animal experimentation and care of animals under study. Experiments done in accordance with the Danish legislation on animal husbandry do not require further ethical permissions.

2.2. Experimental Design

The study was designed to compare the effect of three different curative treatments on tail damage and behaviour in pens with a tail biting outbreak. On the day of the tail biting outbreak (defined as day 0; see Section 2.4 for outbreak definition), one of three treatments were randomly allocated to the pen: straw, rope, or Bite-Rite. To follow the development in tail injuries, tails were scored once weekly after an outbreak was noted and the behaviour of the pigs was recorded on day −1, 0, 2, and 7.

2.3. Animals and Housing

This study included 1987 undocked DanBred crossbred ((Landrace × Large White) × Duroc) nursery pigs in 65 pens as they grew between approximately 6–30 kg. Pigs originated from four different farrowing batches with 458–525 pigs per batch. Pigs were born in a farrowing system with loose sows (for pen design details, see Pedersen et al. [13]). Iron injections (Uniferon, Pharmacosmos, Holbæk, Denmark), grinding of the needle teeth (Tandsliber proff, Hatting, Horsens, Denmark) and surgical castration of male piglets with a scalpel were carried out on day 3 or day 4 after birth. Male piglets were given analgesic just before castration (Melovem® 5 mg/mL, Raamsdonksveer, The Netherlands). Throughout the lactation period, piglets had access to the straw that the sow pulled from a straw rack. Approximately two weeks after farrowing, piglets were offered solid feed on the floor in the creep area. Two days before weaning, the pigs were ear tagged and their sex were noted. The lactation period was 27.8 days (SD = 2.9) and pigs weighed 5.8 kg (SD = 1.5) at weaning. After weaning, pigs were transported to a nursery facility close to the sow facility (1.5 km).

At the nursery facility, pigs were sorted by size within a batch and randomly allocated to nursery pens with 30 pigs/pen (SD = 2; 0.35 m^2/pig). Sex distribution was on average 49.2% (range = 32.2% to 77.4%) gilts per pen. The four experimental rooms consisted of 26 or 30 pens, and between 18–20 pens per room were included in the experiment. Pens measured 4.85 × 2.18 m (length × width) with 7.1 m^2 solid floor and 3.5 m^2 cast iron slatted floor. A 2.16 m^2 adjustable covering was placed above the solid floor in the lying area. Adjacent pens shared a dry feed dispenser with two nipple drinkers, one placed on each side of the feed dispenser (MaxiMat Weaner 7–60 kg, Skiold A/S, Sæby, Denmark). Additionally, a separate water supply (drinking bowl) was placed next to the feed dispenser towards the slatted floor. Each pen was equipped with two wooden blocks hanging from a chain just above the floor without touching the floor to ensure permanent access to enrichment according to legislation [14]. Pens were provided daily with approximately 350 g of finely chopped straw (Easy Strø, Easy Agri Care, Nykøbing Mors, Denmark) on the solid floor. Artificial lighting was on from 06:00 h–22:00 h.

The rooms were ventilated by a negative pressure air flow through wall air inlets on one side of the building (SKOV A/S, Glyngøre, Denmark). The room temperature at weaning was 24 °C and was gradually lowered to 19 °C on day 42. Thermostatically controlled floor heating pipes were placed inside the floor in the lying area, giving a floor temperature of 30 °C at the start of the study. The floor heating was turned off on day 14.

Pigs were fed three different home-mixed compound diets (ad libitum access) from 6–30 kg. The diets were formulated to fulfil the nutritional requirements of pigs of this age and genotype. Phase 1 diet allocated from approximately 6–10 kg (17.4% crude protein) consisted of 64.0% wheat, 22.0% premix of minerals and vitamins (HeavyPig 3 20%, Vilomix, Mørke, Denmark), 10.5% fish meal, and 3.5% soy oil. Phase 2 diet allocated from approximately 10–15 kg (18.1% crude protein) consisted of 44.4% wheat, 25.0% barley, 15.0% toasted soy bean, 8.2% premix of mineral and vitamins (MIN 27600, Vilomix, Mørke, Denmark), 5.0% fish meal, and 2.4% soy oil. Phase 3 diet allocated from approximately 15–30 kg (18.4% crude protein) consisted of 48.8% wheat, 25.5% toasted soy bean, 20.0% barley, 4.2% premix of mineral and vitamins (MIN 27603, Vilomix, Mørke, Denmark), and 1.5% soy oil. The transition between feed compounds depended on the average body weight in the pen at weaning and was gradually conducted over a 7- or 14-day period. Pigs were housed at the nursery location for 6.5 weeks before being moved to the finisher facility, at which point they left the study.

During the stock person's daily inspection, pigs with clinical signs of disease were treated with antibiotics when needed according to the herd veterinarian's recommendations. Unthrifty animals and pigs with severe tail lesions (more than half the tail missing or swelling as a sign of infection) were moved to hospital pens.

2.4. Tail Scoring and Tail Biting Outbreak (Day 0)

From weaning until a tail biting outbreak, tails were scored three times weekly (Monday, Wednesday, and Friday) to determine the day of the outbreak. Tail damage severity and tail posture were recorded in the same way as in Lahrmann et al. [12] by researchers. During tail scoring, ear tag number, tail damage severity, wound freshness, tail length, and tail swelling on injured tails were recorded on each individual according to the criteria listed in Table 1.

Table 1. Tail injury classification after Lahrmann et al. [12].

Tail Scoring	Description
Damage severity	
No	No visible tail lesions. The earlier lesion is healed.
Minor scratches	Minor superficial scratches.
Wound	Visible wound and tissue damage larger than a few millimetres in diameter.
Wound—tail end will fall off	The outer part of the tail has almost been bitten off. During healing, the tail tip will fall off.
Damage freshness	
Intact scab	The wound is covered with a hard, dry scab.
Not intact scab	The wound is covered with a scab, but cracks in the scab and dried blood/fresh tissue are visible.
Fresh wound—not bleeding (weeping)	Skin is broken, no scab, no blood—only weeping.
Fresh wound—bleeding	Fresh lesion and fresh blood are visible.
Tail length	
Intact	Full-length tail.
Outer part is missing	The outer part of the tail is missing.
More than half of the tail is missing	More than half of the tail is missing.
<1 cm left of the tail	Less than 1 cm of the tail is left.
Swelling	
No swelling	No swelling.
Swelling is present	Swollen red tail indicating an infection.

A "tail biting outbreak" occurred when at least four pigs in a pen (approximately 13% of the pigs) had a tail wound (see definition of a wound in Table 1) irrespective of "damage freshness", "tail length", and "swelling". A wound was more severe than "minor scratches". The day of the tail biting outbreak was set as day 0. The stock person did not observe any tail biting outbreaks during the daily inspection of pigs between recording days.

In pens with a tail biting outbreak, tails were scored once weekly on day 7, day 14, day 21 and so on after the outbreak. Tail scoring continued, until an escalation in tail damaged pigs was observed (see definition in the curative treatment paragraph below).

2.5. Treatments

One of three curative treatments selected in a predetermined random order, was allocated to the pen on the day of the tail biting outbreak (day 0): straw (23 pens), rope (22 pens) or a commercially available hanging plastic enrichment with four chewable rods (Bite-Rite, Ikadan Systems A/S, Ikast, Denmark, 20 pens). A curative treatment was provided in 65 pens and it was maintained until the pigs were moved to the finisher facility (study end).

In pens with straw, approximately 200 g (7 g/pig/day) of chopped wheat straw (chopped during harvest in a combine harvester) were provided on the solid floor once a day. This was in addition to the 350 g of finely chopped straw, which all pens received daily throughout the study period.

In pens with rope, a coil of sisal rope (20 mm in diameter) was placed above the pen. The rope was pulled from the coil leaving roughly 30 cm of rope on the solid floor, and the top end at the coil was secured, ensuring that no more rope could be pulled out by the pigs. The rope was provided in the middle of the pen, approximately one meter from the slatted floor. A knot was tied about 20 cm from the rope end to reduce consumption. If the rope end was consumed, the knot was loosened, and new rope was provided the following day in the same way as described above.

In pens with a Bite-Rite, the device was suspended in the middle of the pen at the same location as rope. The plastic rods were located at a height at which pigs could easily reach and chew on them—standing or sitting. As the pigs grew, the Bite-Rite was gradually raised.

2.6. Defining Escalation in Tail Damage and Removing Biters and Victims

Two criteria were used to determine if the curative treatment failed to stop the tail biting behaviour; an escalation in fresh wounds (direct measure) and removing a biter (indirect measure). Fresh wounds were tail damage more severe than "minor scratches" and a fresh wound was either weeping or bleeding according to "damage freshness" (see description in Table 1). This definition of an escalation was chosen, as we wanted a definition where it was clear that the tail biting was ongoing. Removing the biting pig was used as a criterion as this indirect measure also reflected whether the curative treatment served the primary purpose; to stop the tail biting behaviour. When the term "an escalation in tail damage" is applied in this paper, it refers to the sum of pens in which a biter was removed, and pens observed with four fresh wounds or more, either of which imply a failure of the curative treatment to stop the tail biting behaviour. Pens observed with four fresh wounds or pens from which a biter were removed were excluded from the study at this point. In these pens, extra steps were taken to stop the biting behaviour to ensure the welfare of the remaining pigs, such as providing other or more enrichment material.

The four fresh tail wounds and biters could either be observed during weekly tail inspection days (see Section 2.5) or during the daily health inspection. Biters were pigs observed repeatedly chewing/biting the tail so hard that the receiver screamed and reacted by suddenly moving away or turning against the biting pig. Pigs performing this kind of behaviour were removed from the pen.

Tail biting victims with severe tail lesions defined as more than half the tail missing or swelling as a sign of infection were moved to a hospital pen for appropriate veterinary treatment and monitoring.

2.7. Behavioural Recordings on Day −1, 0, 2, and 7

An overhead surveillance video camera (Dahua 2MP HD IR Dome, Dahua, Haarlemmermeer, The Netherlands) was installed above each pen timed to record from 07:00 h to 21:00 h. The time was chosen based on a previous study showing that pigs are most active in the day time [15]. Video data were collected from when the pigs entered the pen and ended when data collection stopped due an escalation in tail injuries or when pigs were moved to the finisher unit.

Behavioural observations were made, using video recordings on day −1 (the day before a tail biting outbreak and introduction of the enrichment treatment), day 0, day 2, and day 7. Not all 65 pens could be included in the behavioural study, either due to poor video quality or due to an escalation in tail biting during the sampling period. As such, behavioural observations were conducted using video from 33 pens (straw $n = 11$, rope $n = 14$, and Bite-Rite $n = 8$).

The behavioural observations were divided in two parts. In the first part, the number of pigs engaged in object interaction or other behaviours (as listed in Table 2) performed by standing, sitting, or lying pigs were recorded in the pen using scan sampling once every 30 min between 07:00 h and 21:00 h. In the second part using continuous sampling, the frequency of tail directed behaviour (TDB) was recorded at the pen level as listed in Table 3 during a 10-min period every 2nd hour between 07:00 h and 21:00 h, starting at 07:00 h. Tail interest (TI) was recorded if it lasted more than one second, and if the pig repeated the behaviour after a pause of 2–3 s, it was recorded as a new incidence. Additionally, if the pig shifted between two types of tail directed behaviour towards the same recipient, both behaviours were recorded, for example: tail interest, to tail-in-mouth, further to two-stage tail biting, and back to tail interest after the recipient pig reacted to tail biting; a total of four incidences were recorded.

Table 2. Ethogram describing recorded pigs' behaviours using scan sampling. Based on by Day et al. [16] and Nannoni et al. [17].

Behaviours	Description
Walking	Standing up on all four legs and showing a clear walking pattern.
Sitting active	The rear end is planted on the floor and the front legs are straight. The pig is actively engaged in one of the behaviours described below.
Lying active	Resting ventrally or laterally on the floor. The pig is actively engaged in one of the behaviours described below.
Object interaction	The head is oriented towards the object and the head is not more than a pig's length from the object expressing interest. The pig might also manipulate the enrichment with its mouth or snout or perform rooting behaviour.
Wooden stick/chain	Manipulating the wooden sticks or chain with the snout or mouth or using the wooden sticks for other physical purposes like scratching.
Exploring the floor	The snout was touching the floor. The pig performed circling or back and forth head movements or chewing substrate lying on the floor (rooting behaviour).
Other pen objects	Interaction with other pen objects like pen equipment (biting bars, feeder etc.).
Pen mate directed behaviour	Behaviours directed towards pen mates and pigs from the neighbouring pen through the bars. Those behaviours were: snout contact, chasing, headbutting, nosing, snapping at body parts except for the tails, or fighting.

Table 3. Ethogram describing tail directed behaviours recorded during continuous sampling. Based on Taylor et al. [18] and Zonderland et al. [19].

Tail Directed Behaviour (TDB)	Description
Tail interest (TI)	The performing pig's snout was close to and fixated to the recipient's rear end and making nosing movements.
Tail-in-mouth (TIM)	The performing pig had the recipient's tail in its mouth or was chewing on it without a response from the recipient.
Two-stage tail biting (TB)	The performing pig had the recipient's tail in its mouth and was visibly pulling the tail, or the recipient responded to the tail biting with abrupt movements such as (but not exclusively): jumping, turning, running away, changing resting position, and sudden head movement.
Sudden forceful tail biting (TB)	The performing pig was striking out after another pig's tail suddenly and forcefully. The tail biting was performed without any earlier tail directed sequences such as tail interest. The recipient pig showed similar response as to two-stage tail biting.

2.8. Statistical Analysis

Statistical analyses were performed in SAS Enterprise Guide 7.1 (SAS Institute Inc., Cary, NC, USA) with a significance level of $p < 0.05$.

2.8.1. Escalation in Tail Injuries

An intervention was conducted in 65 pens, but in four pens, the tail biting outbreak occurred within the last week of the experiment (straw $n = 1$, rope $n = 2$ and Bite-Rite $n = 1$). These four pens were excluded from the analyses because the effect of the curative treatment could only be tested for less than one week.

The difference in the number of tail damaged pigs between treatments on day 0 and the effect of the number of tail damaged pigs on day 0 on the risk of an escalation in tail damage was analysed using the general linear mixed procedure (GLIMMIX). The treatment and days after weaning until outbreak were fitted as fixed effects and batch as random effect.

The GLIMMIX procedure was also used to analyse the effect of treatment (straw $n = 22$, rope $n = 20$ or Bite-Rite $n = 19$) at pen level on a potential escalation (binary outcome) in tail damage (removing biter or at least four fresh wounds). In this analysis, treatment and days until outbreak were fitted as fixed effect and batch as random effect.

2.8.2. Tail Damage Severity

Tail injuries were grouped according to severity and tail length, but irrespective of damage freshness (0 = no tail damage, 1 = tail damage and full tail length (mild), 2 = tail damage and tail loss or swollen tail (moderate)) before statistical analysis. The difference in the number of tail damaged pigs between treatments on day 7 and day 14 in pens ($n = 27$) without an escalation in tail injuries was analysed using the GLIMMIX procedure. The treatment, the days after weaning until outbreak, and tail damaged pigs on day 0 (and day 7 in the model analysing differences on day 14) were fitted as systematic effects and batch as random effect in the model.

The percentage of pigs with tail damage (category 1 or 2) on day 0, day 7, and day 14 in pens without a subsequent escalation in tail injuries irrespective of treatment and that had been at the facility for at least 14 days after treatment start were analysed ($n = 27$) using the mixed linear procedure (MIXED) with days after weaning until outbreak and day after intervention (day 0, 7, or 14) as fixed effect. Batch and pen were included as random effects.

2.8.3. Behavioural Observations

Behavioural recordings were conducted in 33 pens. Some video sequences were disturbed or not available. Missing sequences accounted for 1.8% of the total planned scans and 2.6% for continuous sampling.

The percentage of active pigs was calculated as the proportion of the total number of pigs in the pen. The other behaviours recorded using scan sampling were calculated as the proportion of active pigs (sum of walking, standing, sitting active, and lying active). The proportion of pigs engaged in hanging object interaction was analysed for the Bite-Rite and rope treatment. In pens with the straw treatment, it was not possible to distinguish between explorative behaviour directed against the straw and floor. Therefore, in order to compare the three treatments, exploratory behaviour towards the floor, substrate and hanging object were summed. Behavioural recordings collected by scan sampling were analysed using the MIXED procedure. No interaction between treatment and day was present.

Tail directed behaviour (TI, tail-in-mouth (TIM), and tail biting (TB)) were analysed as mean frequencies per day within the observation period (80 min per day). Due to very low frequencies, sudden-forceful-biting, and two-stage-biting were grouped together. Tail directed behaviour (continuous sampling) was analysed using the GLIMMIX procedure with day and treatment (straw $n = 11$, rope $n = 14$, Bite-Rite $n = 8$) as fixed effects and pen × treatment and batch as random effects. Tail directed behaviour estimates were back transformed using the i-link function. The analysis demonstrated an interaction between treatment and day.

Results from analysis in the GLIMMIX and MIXED procedure are presented as least square means (LSmean) and standard error (SE).

3. Results

On average, a tail biting outbreak developed 25 days (SD = 10.2; range = 9–45 days; median = 23 days) after weaning (Figure 1), and 6.8 pigs/pen (SD = 3.4; range = 4–21 pigs) had tail damage at that timepoint (day 0). An escalation in tail damage (four fresh tail wounds or removing a biter; $n = 25$) occurred on average 14 days (SD = 9.2; range = 2–34 days) after the outbreak. See Table 4 with descriptive results presented by treatment. In four pens of the 25 pens with an escalation in tail damage, removing the biter was the reason, whereas in 21 pens, the incidence of four fresh wounds was the cause (Table 4).

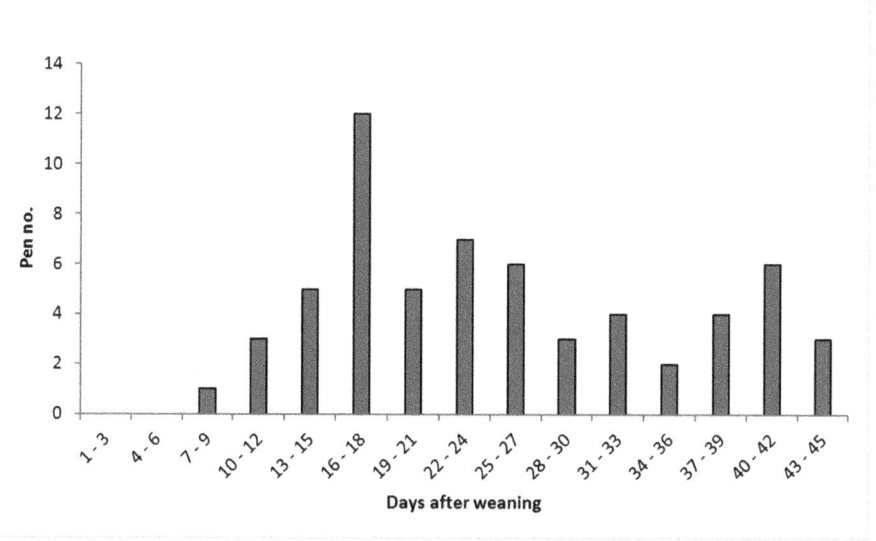

Figure 1. Incidence of first tail biting outbreak at pen level during the study period ($n = 61$).

Table 4. Number of pens, days until outbreak, tail damaged pigs on day 0, number of pens with an escalation in tail damage, and days until escalation presented within treatment.

	Treatment		
	Straw	Rope	Bite-Rite
Pens with curative treatment, n	22	20	19
Days till tail biting outbreak [1]	22 (10.7)	22 (10.0)	20 (10.1)
Tail damaged pigs on day 0, n [1,2]	6.4 (3.1)	6.7 (2.7)	7.2 (4.4)
Pens with an escalation in tail damage, n			
four fresh wounds	4	6	11
removed biter	2	1	1
Days till an escalation in tail damage [1]	21 (11.3)	11 (4.8)	12 (8.9)

[1] Results are presented as mean (SD). [2] Pigs within pen with at least a tail wound. Minor superficial scratches were not encountered as tail damage in the definition of a tail biting outbreak.

3.1. Effect of Curative Treatments on Escalation in Tail Damage

In total, 61 pens were included in the statistical analysis (straw $n = 22$; rope $n = 20$; Bite-Rite $n = 19$). On day 0, when one of the three curative treatments were provided, the number of tail damaged pigs at pen level did not differ between treatments ($F_{2,55} = 0.19$, $p = 0.83$) and the number of tail damaged pigs on day 0 did not affect the risk of a subsequent escalation in tail damage ($F_{1,55} = 0.10$, $p = 0.76$).

An escalation in tail damage (sum of removed biting pig and pens with four fresh wounds) occurred in more pens with Bite-Rite than in pens with straw (Figure 2). There were no significant differences between rope and either Bite-Rite or straw.

In total, fewer pigs received tail damage in straw pens (16.7%) compared to Bite-Rite (25.6%; $t_{(1811)} = 3.81$, $p < 0.001$) and rope (22.8%; $t_{(1811)} = 2.69$, $p < 0.01$). Whereas no difference in tail damaged pigs was observed between rope and Bite-Rite pens ($t_{(1811)} = 1.12$, $p = 0.26$).

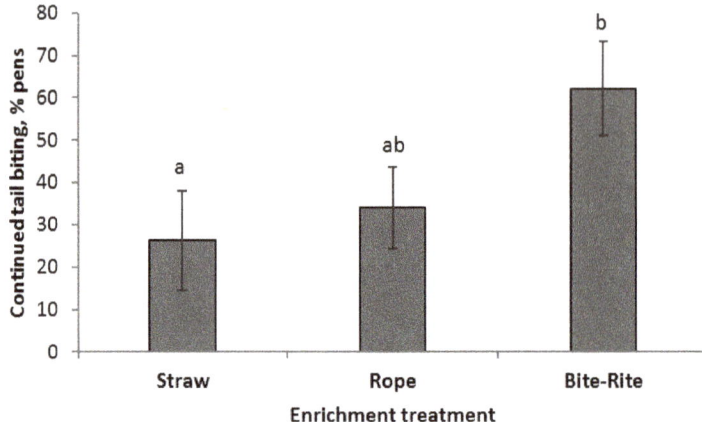

Figure 2. Percentage of pens with an escalation in tail damage after the initial outbreak presented as least squares mean (LSmean) (±standard error (SE)) according to treatment. Different superscript (a, b) indicates a significant difference of $p < 0.05$ between treatments.

3.2. Effect of Curative Treatments on Behaviour

There was no difference in activity between treatments, but pigs were more active on the day of the outbreak (day 0) than on the other days (Table 5). More behaviour was directed towards the rope than the Bite-Rite, and more behaviour was directed against the objects on day 0 than on day 2 and day 7 (Table 5). More exploratory behaviour was observed in pens with rope than in pens with Bite-Rite and straw, and more exploratory behaviour was observed on day 0 than on day −1, 2, and 7 (Table 5). Furthermore, the level of pen mate directed behaviour (excluding tail-directed behaviour) was lower on day 0 and 2 than on day −1. No difference in pen mate directed behaviour was observed between treatments.

Table 5. The frequency of behaviours (percentage of pigs) observed on day −1, 0, 2, and 7 and within each treatment (straw, rope and Bite-Rite). Different lowercase letters (a, b and c) indicate significant difference of $p < 0.05$ between days or between enrichment treatment.

Behaviour	Day [1]						Treatment [1]				
	−1	0 [3]	2	7	SE	p-Value	Straw	Rope	Bite-Rite	SE	p-Value
Active pigs, %	37.0a	39.6b	37.4a	36.8a	0.81	<0.01	38.3	36.7	38.1	1.21	0.49
Object interaction, %	-	23.3a	12.7b	10.2b	1.40	<0.001	-	18.9a	11.9b	1.32	<0.001
Explorative behaviour, %	29.7a	40.5b	36.8c	35.1c	1.03	<0.001	32.7b	39.0a	34.8b	1.34	<0.001
Pen mate directed behaviour except TDB, % [2]	12.3a	8.26b	9.55c	11.4a	0.56	<0.001	11.1	9.7	10.3	0.84	0.35

[1] The sum of the recorded behaviours does not sum to 100%, as pigs could be engaged in other behaviours aside from those presented in the table. [2] TDB; tail directed behaviour. [3] Day 0; the day of the tail biting outbreak.

Figure 3 shows the effect of enrichment treatment on tail interest (TI), tail-in-mouth (TIM), and tail biting (TB) on day −1, 0, 2, and 7. Irrespective of enrichment treatment TI, TIM, and TB were lower on day 0 than on day −1. Furthermore, except for TIM on day 7 in Bite-Rite pens, the frequency of TI, TIM, and TB was lower on day 2 and 7 than on day −1. In pens with Bite-Rite, the frequency of TI, TIM, and TB was higher on day 7 than on day 0 and 2. There was no difference in TIM and TB in rope and straw pens on day 7, day 0, or day 2, but TI was observed more often in rope and straw pens on day 7 than on day 2, but not on day 0.

TB differed between treatments on day 0 with a lower prevalence in pens with Bite-Rite than in pens with rope and straw, and TIM was lower in pens with Bite-Rite than in pens with rope on day 0.

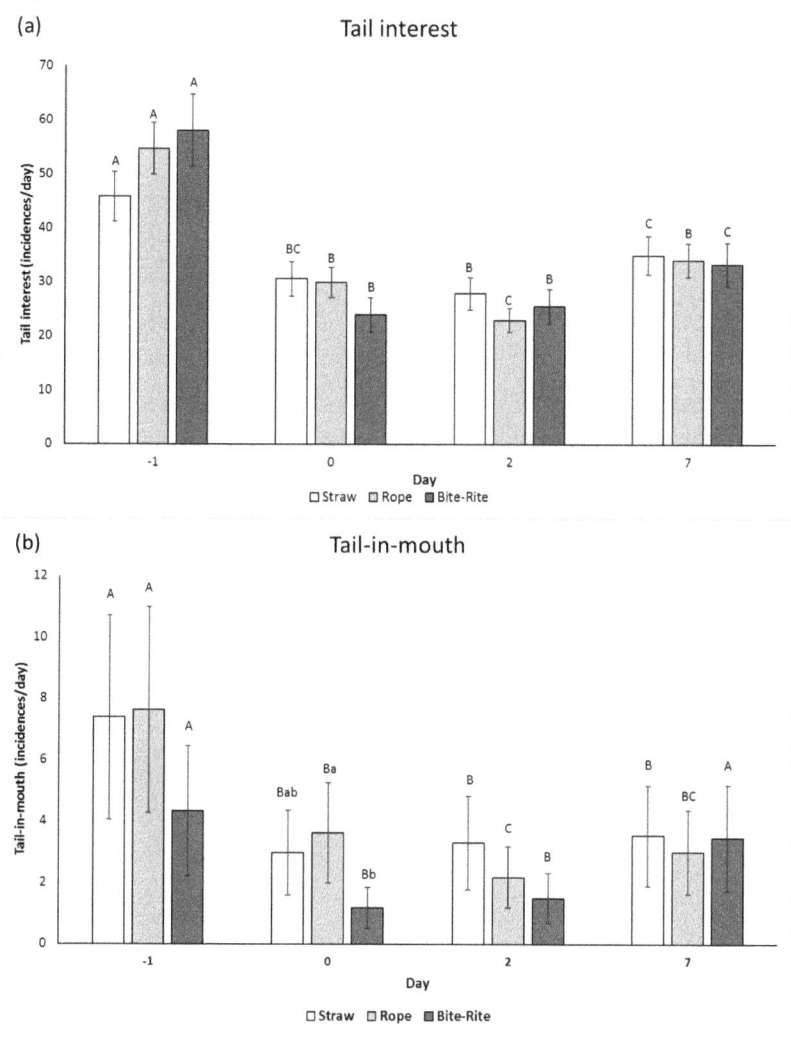

Figure 3. Cont.

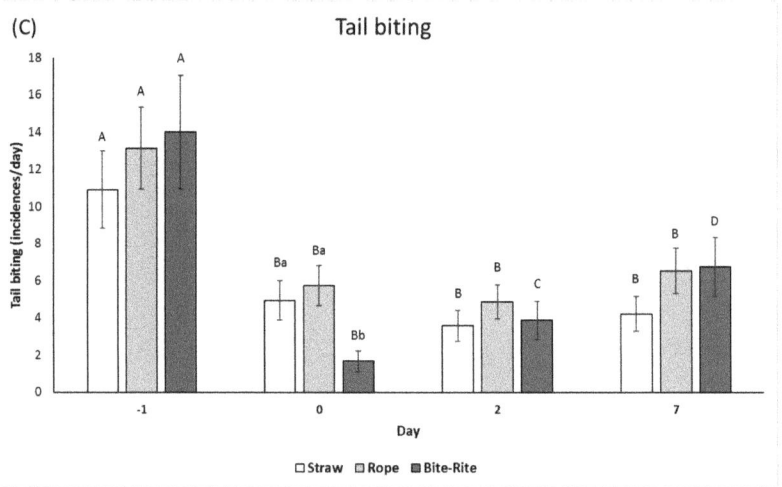

Figure 3. Tail interest (**a**), tail-in-mouth (**b**), and tail biting (**c**) presented as frequency per day (LSmean ± SE) for each enrichment treatment (straw, rope and Bite-Rite) on day −1, 0, 2, and 7 (80 min observation period per day). Different lowercase letters indicate a significant difference of $p < 0.05$ between treatments within days. Different uppercase letters indicate a significant difference of $p < 0.05$ between days within treatments.

3.3. Tail Damaged Pigs on Day 0, 7, and 14 after Curative Treatment in Pens without an Escalation

No difference in the number of tail damaged pigs was observed between treatments on day 7 ($F_{2,21} = 0.08$, $p = 0.92$) or on day 14 ($F_{2,20} = 1.0$, $p = 0.37$) after the first outbreak in pens without an escalation in tail damage. Combining data from the three treatments with no subsequent escalation in tail damage (n = 27 (straw $n = 11$, rope $n = 11$ and Bite-Rite $n = 5$)), 20.7% of the pigs per pen (6.4 pigs per pen; SD = 3.5) had tail damage on day 0. In these pens, fewer pigs had tail damage on day 14 (5.8%; 1.7 pigs per pen; SD = 2.1) than on day 0 and day 7 (17.1%; 5.0 pigs per pen; SD = 3.5), but there was no difference between day 0 and day 7 (Figure 4).

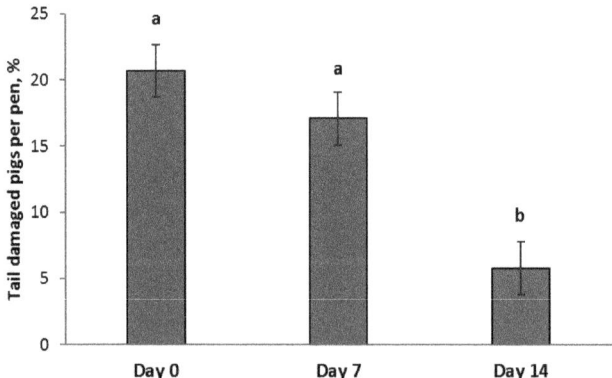

Figure 4. Percentage of tail damaged pigs per pen on day 0, day 7, and day 14 (LSmean ± SE) after the intervention in pens without an escalation in tail damage (n = 27). Different superscripts (a, b) indicate significant difference of $p < 0.001$.

4. Discussion

This is, to our knowledge, the first study to compare the effect of straw, rope, and Bite-Rite as curative treatments during an ongoing tail biting outbreak (see also Lahrmann et al [20]). An escalation in tail damage was observed in 26% of the straw pens, 34% of the rope pens, and 62% of the Bite-Rite pens. Providing a small amount of straw on the floor during an ongoing tail biting outbreak reduced the risk of a further escalation in tail damage more efficiently than providing a Bite-Rite. Numerically, rope fell between the other two treatments, but no significant difference was demonstrated between pens provided with rope and pens provided with straw or Bite-Rite.

On average, a tail biting outbreak occurred 23 days after weaning (range = 5–45 days), which is in line with Zonderland et al. [8] reporting tail biting outbreaks in 50% of the pens with undocked tails (median) 24 days after weaning (range = 8–31 days). In agreement, Veit et al. [21] reported that tail damage began to occur two to three weeks after weaning in pens with undocked tails, while D'Eath et al. [22] reported a mean outbreak day of 28 days post-weaning (range = 16–41 days).

In pens with rope, pigs interacted significantly more with the object than in pens with Bite-Rite. These findings are in line with a review by Studnitz et al. [23] concluding that pigs' preference towards an object will increase with its destructibility and complexity. As the rope end laid on the floor, the material might have been available to more pigs and the material was more destructible than the hanging Bite-Rite. Both in rope and Bite-Rite pens, a decline in object interaction was observed on day 2 after the intervention. This indicates that the effect of novelty decreased within a few days. Other studies including a Bite-Rite demonstrated similar results with a higher level of enrichment manipulation in pens with straw than in pens with Bite-Rite [24,25]. In pens with rope, pigs performed more exploratory behaviour than in straw and Bite-Rite pens. The provided straw gradually disappeared over time and was therefore not permanently present, which could explain the higher level of exploratory behaviour in pens with permanent access to rope on the floor.

Irrespective of treatment tail interest, tail-in-mouth and tail biting declined from day −1 to day 0 and day 2. Bite-Rite caused the greatest reduction in tail biting from day −1 to day 0, compared to the other two treatments. However, each of the tail-directed behaviours increased during the following seven days in Bite-Rite pens, while the frequency remained more constant in pens with rope or straw. It may indicate that the attractiveness of Bite-Rite was higher at first, but faded quicker than the other two materials, which gave rise to more tail-directed behaviour on day 7. This is in line with a study by Van de Weerd et al. [25] reporting more tail biting in pens with a Bite-Rite than in pens with straw bedding.

In pens without a further escalation in tail damage, fewer pigs had tail damage on day 14 than on day 0 and day 7. This suggests that it took more than seven days for a tail wound to heal with the allocated enrichment materials. These results are supported by Lahrmann et al. [26] reporting that 89% of the tail wounds healed within 14 days. However, as discussed by Lahrmann et al. [26], the duration of the wound healing is undoubtedly affected by the severity of the wound.

In the present study, even if the most efficient curative treatment (7 g/pig/day of chopped straw on the pen floor) was applied, the tail damage escalated in approximately one in four pens. A different intervention strategy or other kinds of interventions are likely needed to stop the biting behaviour even more efficiently during a tail biting outbreak. These interventions could include removing tail damaged pigs, providing the materials used in the present study in combination, providing materials in a greater quantity, providing other types of enrichment or perhaps shifting between enrichment materials during the post-outbreak phase. In a study by Zonderland et al. [8], an allocation of straw twice daily (20 g/pig/day) or removing the biting pig as curative treatments were equally likely to stop the tail biting. In the Zonderland study, the treatment effect was measured as the prevalence of pigs with a fresh wound in the following ten days after the tail biting outbreak. However, ten days after the intervention, 11% of the pigs still had a fresh tail wound, compared with 25% on the day of the outbreak, suggesting that even these interventions did not stop the biting behaviour completely.

The curative treatments applied in the present study were chosen due to their feasibility in current production systems. In the present study, straw as curative treatment was given additionally to very finely chopped straw provided on a daily basis. It could be that providing a larger amount of straw and thereby ensuring access for a longer period of the day would have reduced the escalation in tail biting more efficiently [27]. Hence, increasing amounts of straw, increases the time pigs interact with the material [28,29]. There are, however, practical problems with larger amounts of straw, as it increases the risk of the material accumulating in the slurry canals or blocking up the slurry pipes [9]. To ensure access to straw for a longer period during the day, Oxholm et al. [28] demonstrated that more frequent allocation (four times daily vs. once a day) of the same total amount of straw ensured more straw left in the pen the following day. Another approach could be to give the straw in a rack, which would probably also increase accessibility. However, straw in a rack might not be as effective as straw on the floor, if pigs have difficulty pulling the straw from the rack and on to the floor [8].

Another possibility could be to allocate a material that pigs find more attractive than straw. In a review by Studnitz et al. [23], more complex materials (mixtures) were ranked higher than straw, when evaluated according to attractiveness. These materials might stop tail biting more efficiently, because pigs find them more attractive even in smaller amounts. However, tail biting studies like the current one are required to establish the materials' effect during a tail biting outbreak.

In pens provided with rope, a knot was tied approximately 20 cm from the rope end. The rope treatment might have prevented an escalation in tail damage to a greater extent without the knot. When rope beneath the knot was consumed, the rope was no longer destructible and chewable. This may have caused less interaction with the material, until new rope was released the following day. It may also be that several pieces of rope are needed to stop the biting completely. This would give more pigs access to the material simultaneously and may avoid pigs experiencing frustration due to lack of access to the material, as discussed by Docking et al. [30]. However, in a minor study by Scott et al. [31] (eight pens per treatment), access to four toys vs. one toy made of alkathane piping did not increase the proportion of observations at which pigs interacted with the material.

Novelty, besides destructibility and manipulability, is an important feature to keep pigs interested [23]. Therefore, a shift between different types of hanging materials could have been more effective in reducing tail biting. Perre et al. [32] reported in a minor study (six pens per treatment) that shifting between hanging enrichment objects reduced tail damage and biting behaviour compared to only providing a chain.

Tail biting occurs sporadically [9], which makes the planning and control of such a study difficult. In addition, while studies have to be scientifically sound, studies of injurious behaviour such as tail biting, have to balance the need for information with the welfare of the animals. In the present study definitions of a tail biting outbreak and when to remove a biting pig, had to be decided based on these two factors.

5. Conclusions

Providing additional straw in a relatively small amount (7 g/pig/day) on the floor during a tail biting outbreak reduced the risk of an escalation in tail damage more effectively than providing a Bite-Rite, while rope reduced tail damage at an intermediate level which was not significantly different from either straw or Bite-Rite. All three treatments reduced tail directed behaviour on the day of the outbreak compared to the previous day. However, during the seven days recording period after an outbreak, the level of tail directed behaviour increased in pens with Bite-Rite. This increase in tail directed behaviour did not occur in pens with rope and straw. Pigs also interacted more with rope than with Bite-Rite. In general, the results indicate that a Bite-Rite cannot keep pigs interested for very long and it should be combined or rotated with other materials to successfully stop tail biting.

The curative treatments applied in the study were chosen due to their ability to be used under commercial conditions. However, tail damage escalated in approximately one in four straw pens. This indicates that other enrichment treatments or different interventions strategies are needed to more efficiently stop the tail biting behaviour. Future studies comparing the effect of different interventions strategies during a tail biting outbreak are needed to establish this.

Author Contributions: Conceptualization, H.P.L.; methodology, H.P.L., J.F.F., B.F., R.D'E., and C.F.H.; formal analysis, H.P.L.; investigation, H.P.L. and J.F.F.; data curation, H.P.L.; writing—original draft preparation, H.P.L. and J.F.F.; writing—review and editing, C.F.H., R.D'E., J.P.N., B.F.; supervision, C.F.H., R.D'E., J.P.N., B.F.; project administration, H.P.L.; funding acquisition, H.P.L., C.F.H., and B.F.

Funding: This work was supported by The Danish Pig Levy Foundation and Innovation Fund Denmark (4135-00081B). The work of SRUC was supported by the Rural and Environment Science and Analytical Services Division of the Scottish Government.

Acknowledgments: The authors would like to acknowledge pig producer Niels Aage Arve and his staff for making his herd available for the study. Further, we would like to thank Hans Peter Thomsen and Mimi Lykke Eriksen from SEGES, Danish Pig Research Centre for technical assistance and Mai Britt Friis Nielsen from SEGES, Danish Pig Research Centre for statistical help.

Conflicts of Interest: The authors declare no conflict of interest. The funders had no role in the design of the study; in the collection, analyses, or interpretation of data; in the writing of the manuscript, or in the decision to publish the results.

References

1. Alban, L.; Petersen, J.V.; Busch, M.E. A comparison between lesions found during meat inspection of finishing pigs raised under organic/free-range conditions and conventional, indoor conditions. *Porc. Health Manag.* **2015**, *1*, 1–11. [CrossRef] [PubMed]
2. Kongsted, H.; Sørensen, J.T. Lesions found at routine meat inspection on finishing pigs are associated with production system. *Vet. J.* **2017**, *223*, 21–26. [CrossRef] [PubMed]
3. Edwards, S. What to do about tail biting today? *Pig J.* **2011**, *66*, 81–86.
4. EFSA. Scientific Report on the risks associated with tail biting in pigs and possible means to reduce the need for tail docking considering the different housing and husbandry systems (question no. EFSA-Q-2006-013). *EFSA J.* **2007**, *5*, 1–13. Available online: http://www.efsa.europa.eu/en/efsajournal/doc/611.pdf (accessed on 7 April 2014).
5. Valros, A.; Ahlström, S.; Rintala, H.; Häkkinen, T.; Saloniemi, H. The prevalence of tail damage in slaughter pigs in Finland and associations to carcass condemnations. *Acta Agri. Scan.* **2004**, *54*, 213–219. [CrossRef]
6. Ursinus, W.W.; Wijnen, H.J.; Bartels, A.C.; Dijvesteijn, N.; van Reenen, C.G.; Bolhuis, J.E. Damaging biting behaviors in intensively kept rearing gilts: The effect of jute sacks and relations with production characteristics. *J. Anim. Sci.* **2014**, *92*, 5193–5202. [CrossRef] [PubMed]
7. Larsen, M.L.V.; Andersen, H.M.-L.; Pedersen, L.J. Which is the most preventive measure against tail damage in finisher pigs: Tail docking, straw provision or lowered stocking density? *Animal* **2017**, *2*, 1–8. [CrossRef]
8. Zonderland, J.J.; Wolthuis-Fillerup, M.; Reenen, C.G.V.; Bracke, M.B.M.; Kemp, B.; den Hartog, L.A.; Spoolder, H.A.M. Prevention and treatment of tail biting in weaned piglets. *Appl. Anim. Behav. Sci.* **2008**, *110*, 269–281. [CrossRef]
9. D'Eath, R.B.; Arnott, G.S.; Turner, P.; Jensen, T.; Lahrmann, H.P.; Busch, M.E.; Niemi, J.K.; Lawrence, A.B.; Sandoe, P. Injurious tail biting in pigs: How can it be controlled in existing systems without tail docking? *Animal* **2014**, *8*, 1479–1497.
10. Brunberg, E.I.; Rodenburg, T.B.; Rydhmer, L.; Kjaer, J.B.; Jensen, P.; Keeling, L.J. Omnivores Going Astray: A Review and New Synthesis of Abnormal Behavior in Pigs and Laying Hens. *Front. Vet. Sci.* **2016**, *3*, 57. [CrossRef]
11. Van de Weerd, H.A.; Docking, C.M.; Day, J.E.L.; Breuer, K.; Edwards, S.A. Effects of species-relevant environmental enrichment on the behaviour and productivity of finishing pigs. *Appl. Anim. Behav. Sci.* **2006**, *99*, 230–247. [CrossRef]
12. Lahrmann, H.P.; Hansen, C.F.; D'Eath, R.; Busch, M.E.; Forkman, B. Tail posture predicts tail biting outbreaks at pen level in weaner pigs. *Appl. Anim. Behav. Sci.* **2018**, *200*, 29–35. [CrossRef]
13. Pedersen, J.H.; Moustsen, V.A.; Nielsen, M.B.F.; Hansen, C.F. Temporary confinement of loose-housed hyperprolific sows reduces piglet mortality. *J. Anim. Sci.* **2015**, *93*, 4079–4088.

14. Danish Government. BEK nr. 17 af 07/01/2016, Bekendtgørelse om Beskyttelse af Svin (Order on Protection of Pigs); Fødevareministeriet (Danish Ministry of Food): Copenhagen, Denmark, 2016. Available online: https://www.retsinformation.dk/pdfPrint.aspx?id=176842 (accessed on 2 June 2019).
15. Beattie, V.E.; O'Connell, N.E. Relation between rooting behaviour and foraging in growing pigs. *Anim. Welf.* **2002**, *11*, 295–303.
16. Day, J.E.L.; Burfoot, A.; Docking, C.M.; Whittaker, X.; Spoolder, H.A.M.; Edwards, S.A. The effects of prior experience of straw and the level of straw provision on the behaviour of growing pigs. *Appl. Anim. Behav. Sci.* **2002**, *76*, 189–202. [CrossRef]
17. Nannoni, E.; Sardi, L.; Vitali, M.; Trevisi, E.; Ferrari, A.; Barone, F.; Bacci, M.L.; Barbieri, S.; Martelli, G. Effects of different enrichment devices on some welfare indicators of post-weaned undocked piglets. *Appl. Anim. Behav. Sci.* **2016**, *184*, 25–34. [CrossRef]
18. Taylor, N.R.; Main, D.C.J.; Mendl, M.; Edwards, S.A. Tail-biting: A new perspective. *Vet. J.* **2010**, *186*, 137–147. [CrossRef] [PubMed]
19. Zonderland, J.J.; Schepers, F.; Bracke, M.B.M.; den Hartog, L.A.; Kemp, B.; Spoolder, H.A.M. Characteristics of biter and victim piglets apparent before a tail-biting outbreak. *Animal* **2011**, *5*, 767–775. [CrossRef]
20. Lahrmann, H.P.; Hansen, C.F.; D'Eath, R.B.; Busch, M.E.; Nielsen, J.P.; Forkman, B. Early intervention with enrichment can prevent tail biting outbreaks in weaner pigs. *Livest. Sci.* **2018**, *212*, 272–277. [CrossRef]
21. Veit, C.; Traulsen, I.; Hasler, M.; Tölle, K.-H.; Burfeind, O.; Beilage, E.g.; Krieter, J. Influence of raw material on the occurrence of tail-biting in undocked pigs. *Livest. Sci.* **2016**, *191*, 125–131. [CrossRef]
22. D'Eath, R.B.; Jack, M.; Futro, A.; Talbot, D.; Zhu, Q.; Barclay, D.; Baxter, E.M. Automatic early warning of tail biting in pigs: 3D cameras can detect lowered tail posture before an outbreak. *PLoS ONE* **2018**, *18*, e0194524. [CrossRef] [PubMed]
23. Studnitz, M.; Jensen, M.B.; Pedersen, L.J. Why do pigs root and in what will they root? A review on the exploratory behaviour of pigs in relation to environmental enrichment. *Appl. Anim. Behav. Sci.* **2007**, *107*, 183–197. [CrossRef]
24. Scott, K.; Taylor, L.; Bhupinder, P.G.; Edwards, S.A. Influence of different types of environmental enrichment on the behaviour of finishing pigs in two different housing systems, 1. Hanging toys versus rootable substrate. *Appl. Anim. Behav. Sci.* **2006**, *99*, 222–229. [CrossRef]
25. Van de Weerd, H.A.; Docking, C.M.; Day, J.E.L.; Edwards, S.A. The development of harmful social behaviour in pigs with intact tails and different enrichment backgrounds in two housing systems. *Anim. Sci.* **2005**, *80*, 289–298. [CrossRef]
26. Lahrmann, H.P.; Busch, M.E.; D'Eath, R.B.; Forkman, B.; Hansen, C.F. More tail lesions among undocked than tail docked pigs in a conventional herd. *Animal* **2017**, *11*, 1825–1831. [CrossRef] [PubMed]
27. Munsterhjelm, C.; Heinonen, M.; Valros, A. Application of the Welfare Quality® animal welfare assessment system in Finnish pig production, part II: Associations between animal-based and environmental measures of welfare. *Anim. Welf.* **2015**, *24*, 161–172. [CrossRef]
28. Oxholm, L.C.; Steinmetz, H.V.; Lahrmann, H.P.; Nielsen, M.B.F.; Amdi, C.; Hansen, C.F. Behaviour of liquid-fed growing pigs provided with straw in various amounts and frequencies. *Animal* **2014**, *8*, 1889–1897. [CrossRef] [PubMed]
29. Jensen, M.B.; Herskin, M.S.; Forkman, B.; Pedersen, L.J. Effect of increasing amounts of straw on pigs' explorative behaviour. *Appl. Anim. Behav. Sci.* **2015**, *171*, 58–63. [CrossRef]
30. Docking, C.M.; Weerd, H.A.V.d.; Day, J.E.L.; Edwards, S.A. The influence of age on the use of potential enrichment objects and synchronisation of behaviour of pigs. *Appl. Anim. Behav. Sci.* **2008**, *110*, 224–257. [CrossRef]
31. Scott, K.; Taylor, L.; Bhupinder, P.G.; Edwards, S.A. Influence of different types of environmental enrichment on the behaviour of finishing pigs in two different housing systems 2. Ratio of pigs to enrichment. *Appl. Anim. Behav. Sci.* **2007**, *105*, 51–58. [CrossRef]
32. Perre, V.v.d.; Driessen, B.; Thielen, J.v.; Verbeke, G.; Geers, R. Comparison of pig behaviour when given a sequence of enrichment objects or a chain continuously. *Anim. Welf.* **2011**, *20*, 641–649.

© 2019 by the authors. Licensee MDPI, Basel, Switzerland. This article is an open access article distributed under the terms and conditions of the Creative Commons Attribution (CC BY) license (http://creativecommons.org/licenses/by/4.0/).

MDPI
St. Alban-Anlage 66
4052 Basel
Switzerland
Tel. +41 61 683 77 34
Fax +41 61 302 89 18
www.mdpi.com

Animals Editorial Office
E-mail: animals@mdpi.com
www.mdpi.com/journal/animals

www.ingramcontent.com/pod-product-compliance
Lightning Source LLC
LaVergne TN
LVHW071952080526
838202LV00064B/6727